RISE OF THE
NECROFAUNA

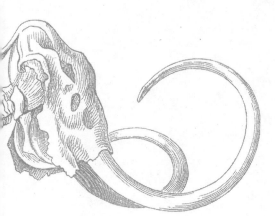

RISE OF THE

BRITT WRAY

FOREWORD BY **GEORGE CHURCH**

The Science, Ethics, and
Risks of De-Extinction

NECROFAUNA

DAVID SUZUKI INSTITUTE

GREYSTONE BOOKS

Vancouver/Berkeley

For Sudan, Najin, and Fatu, the last three (unengineered)
northern white rhinos

Greystone Books Ltd.
www.greystonebooks.com

David Suzuki Institute
www.davidsuzukiinstitute.org

Cataloguing data available from Library and Archives Canada
ISBN 978-1-77164-164-7 (cloth)
ISBN 978-1-77164-163-0 (epub)

Editing by Nancy Flight
Copy editing by Stephanie Fysh
Jacket and text design by Nayeli Jimenez
Jacket illustrations by Brian Tong and iStockphoto.com
Interior illustrations by Iga Kosicka
Printed and bound in Canada on ancient-forest-friendly paper by Friesens

We gratefully acknowledge the support of the Canada Council for the
Arts, the British Columbia Arts Council, the Province of British Columbia
through the Book Publishing Tax Credit, and the Government of Canada
for our publishing activities.

Canadä

CONTENTS

FOREWORD

ONCE UPON A time (in the 1940s), when a person drowned or flatlined from a heart condition, we accepted the outcome as natural and turned our attention to living patients instead. Yet today many people who have drowned or had a seemingly fatal heart attack are saved by CPR and external defibrillators (AEDs). Today we face analogous premature dismissal of dead and dying species. Do we focus our limited resources only on robust populations—or do we embrace rewilding, de-extirpation, and de-extinction of endangered and extinct species?

In practice, we are concerned with ecosystems more than species. But laissez-faire and precautionary principles can be dangerous cop-outs when humans and other species are dealing with rapidly changing environments, and helping healthy species versus challenged or dying ones is not the zero-sum game that is often depicted. Imagine recruiting young volunteers and philanthropists to a cause with a slogan like "We will steadily lose all precious species, but we can delay the inevitable a bit." They would reasonably wander off to more inspiring endeavors. In contrast, if we note successes in reversing such dire trends as the extirpation of the California condor, the bison, and the

American chestnut, then we might get more engagement and a large positive-sum game.

So, do we prioritize what species to save by moral obligation, based on how recently and how big a role our human ancestors played in disrupting those species? More likely, we will focus on practical considerations and the abilities of keystone species and the trophic cascades they produce. For example, six years after the gray wolf was reintroduced in Yosemite, vegetation increased along rivers, soil erosion diminished, and beavers returned, along with their pond geoengineering skills. We have also set as a high priority whales, which provide vital vertical ocean mixing exceeding the sum of physical forces and consequently greatly aiding fixation of carbon. Elephants and their relatives—including the extinct woolly mammoth—routinely knock over large trees, resulting in a lowered ratio of trees to forbs and grasses that aids the reflection of the sun in summer and the penetration of the wind in winter, thereby ameliorating climate change. Elk, deer, and other large herbivores don't typically knock over large trees, but the Zimov team has a Soviet military crawler that simulates the impact of mammoths. It knocks over stumps while hungry herbivores search for grasses beneath the snow, and after a period of heavy grazing the team has observed a 20 degree Celsius (36 degree Fahrenheit) lowering of soil temperature.

In addition to the question of conservation priorities, another issue is the perceived price of "exotic" de-extinctions. The costs of reading and writing DNA in general, and ancient DNA in particular, have dropped about a millionfold so far and continue to drop. These methods do not require frozen samples—a rarity for ancient samples. We already have synthetic genome projects (*E. coli* and yeast) at the 4 million base pair scale—making plausible the 1.4 million changes between Asian elephant and mammoth (or some pragmatic subset of that number). The challenges for de-extinction are likely not the molecular magic but the same as for conventional conservation efforts, such as transforming vegetation and automating social learning—as was done with condor puppets to train the young when parents were inadequate (as may be also true for future revived species, including mammoths). These are just a few of the issues that Wray explores in this timely and thought-provoking book—a beacon of discussion-worthy science at the interface with peculiar policy issues.

GEORGE CHURCH

INTRODUCTION

EARLY ONE MORNING in August 2012, I groggily stumbled to my laptop, coffee mug in hand, and opened my Gmail account to scour the overnight deliveries. The first thing I saw was a newsletter from an online community said to have "the world's smartest website." The site serves as an online storehouse for ideas and conversations of the type shared over dinner parties attended by philosophers, scientists, artists, and public intellectuals, letting those of us who aren't part of their circle get in on their thoughts.

That morning, the subject line read, "To Bring Back the Extinct. A conversation with Ryan Phelan." I stared at those words, cozily sandwiched between an email from my mom and a cheap-flight alert. When I clicked on the link, the piercing blue eyes of a blond, rosy-cheeked, professional-looking woman stared back at me. The words under her image said, "One of the fundamental questions here is, is extinction a good thing? Is it 'nature's way'? And if it's nature's way, who in the world says anyone should go about changing nature's way? If something was meant to go extinct, then who are we to screw around with it and bring it back?"

I slowly brought up more of the page and halted at a specific line of text: "The big question that I'm asking right now is: If we could bring back an extinct species, should we? Could we? How does it benefit society? How does it advance the science?" *Good luck with that,* I thought. But I scanned the interview anyway to try to understand what she was talking about. A few scrolls down, I realized that the idea that she was assembling, and that has gripped me since—the plight to bring extinct species back to life—was not just an important thought experiment but also an experiment with real, breathing bodies behind it and a vital story to tell. This book is an attempt to capture some of the major plot points of that story so far, with an emphasis on the many issues raised by this audacious experiment. But it's worth mentioning that this book does not account for *all* the de-extinction endeavors that have ever taken place. Rather, I focus on the cases that interest me most, which Phelan's words first provoked me to look into.

That interview, as though connected by an IV drip straight into my brain, made my mind churn as the notion of resurrecting extinct species sank in. I leafed through reincarnation tales I'd heard before to try to make some sense of the idea—embalming and burial rituals designed to ward off the body's dilapidation once oxygen ceases to flow, the transhumanist dream to defy death itself and make mortality reversible, the Alcor Life Extension Foundation's offer to swap the blood from your cadaver with antifreeze fluids before cradling your body in a cryogenic vessel (for a nice price, the company will maintain you in a state of suspended animation with the hope of bringing you back to life at a more technologically advanced time). But none of that was what this was about. Humans have wanted to be brought back to life after death for a very long time, yet no one has figured out how to make it happen. So how could Ryan Phelan possibly be proposing it for an extinct species now?

Phelan is a biotech entrepreneur with an eye on the future. She has created several companies, including DNA Direct, the first company to sell genetic tests to consumers. During her time there, she saw the overhead costs drop dramatically for doing lab work with DNA, particularly DNA sequencing—the process in which the order of bases in a DNA molecule is deciphered, allowing for more precise medical diagnostics, forensic ancestry analyses, and so much more. That falling price point got Phelan thinking about what other areas of science might benefit from the decreasing costs of DNA sequencing. And since then, her thinking has helped launch the revival of a longstanding dream in biotechnological circles to resurrect extinct species—a seasoned trope from science fiction.

But in the interview on the website, Phelan didn't refer to "resurrection." Perhaps she was aware of how much trouble that spiritually loaded term could cause in a scientific setting. "We're using the term 'resurgence,' " she said, "because as you can imagine, there's a lot of controversy over if you could bring back an extinct species." She added examples: "Is it invasive? Would it become an invasive species? And is this a bad thing?"

In the years since Phelan spoke those words, a handful of projects aimed at reviving extinct species have gotten off the ground. What were once mild murmurings of species resurgences have become bold headlines about an advancing movement that goes by several names: *resurrection ecology, species revivalism,* and *zombie zoology* are just a few. Related to the idea of zombie zoology is the notion that all of this science might serve only to create *charismatic necrofauna*—a term used by futurist Alex Steffen to describe the cuddly or majestic creatures people might want to see brought back from the dead while preferring that less charismatic specimens stay in their graves. Of all the available labels, *de-extinction* is the one people use most. It is also the one that I rely on in this book. But it is not always clear what de-extinction is, or what good—or bad—it might cause ... not until you look at how different people approach it.

Since 2012, a nonprofit organization called Revive & Restore, which Phelan cofounded with her husband, Stewart Brand, a pioneering environmentalist and technology visionary, has been helping to lay the foundations for de-extinction. The couple has been working hard to turn what was once just a scattering of mad science projects into an emerging field, guided by a concept they call *cautionary vigilance*—a method of analysis they are developing that invites public debate and deliberation of evidence on controversial aspects of innovations. "As opposed to the precautionary principle, which says do nothing because you don't know what the unknown consequences are, we say, use cautionary vigilance with transparency and responsibility," Phelan told me. By biting off only one small chunk of technological risk at a time in the hope of creating a better future, this approach allows Phelan and Brand to iteratively test what is going right and wrong so that they can adapt their projects along the way.

Stewart Brand graduated from Stanford University in 1960 with a degree in conservation biology. During the '60s he published the first edition of the celebrated countercultural *Whole Earth Catalog,* which addressed what the world had coming if it continued on in its unsustainable ways. Up to four times a year between 1968 and 1972, the publication explored a gamut of technologies that could be used in the service of the environmental movement. Steve Jobs called it one of the bibles of his generation, and said "it was sort of like Google in paperback form, 35 years before Google came along: it was idealistic, and overflowing with neat tools and great notions." Its voice was optimistic. The first edition included this bold statement: "We are as gods and might as well get good at it." Today, accusations that genetic engineers and hands-on futurists are guilty of "playing god" because they manipulate nature with technology—as though that's a horrible thing—are a dime a dozen. But the accusation is not something Brand has ever shied away from. He's even updated his catalog's old tagline to say, "We are as gods and *have* to get good at it."

It's a vast understatement to say that after I learned about the work that Phelan and Brand were doing, I became interested in de-extinction. It's more accurate to say that I became transfixed. What started with an email subscription turned into hundreds of hours of reporting and writing: a story for New York Public Radio, then a feature documentary for the Canadian Broadcasting Corporation, and, ultimately, a few years of research that turned into this book. Why? I boil it down to my love of science, which blossomed when I was a biology student in university and became enthusiastic about conservation, a field committed to not only extending but also enriching the lives of other animals. For once, I wasn't doodling in the margins during lectures but fully listening while my favorite professor told us about the experiments he designed to boost populations of dwindling species across southern Ontario. In the final year of my degree, I made a radio documentary for the residents of nearby Opinicon Lake outlining how they could help the local loon population, whose nests were disappearing, do better in the wild. The understanding that "extinction is forever" was writ large for me then. It drove everything that my classmates and I were considering doing with our lives. We were taught to act smartly and swiftly because there will be no second chance.

For better or for worse, I never did become a conservation biologist. Instead, I started a science radio show as a passion project, went to art school, and headed into public broadcasting to make stories for the airwaves. After some years, I enrolled in a PhD program in science communication and moved from my native Canada to Denmark. Although I'm not a traditional journalist, I do documentary work, and it is often about science.

I am still a conservation biology enthusiast and try to keep up with news about the field.

That's how one day, not long after I'd read Phelan's interview, I came across a quotation from an esteemed conservation biologist named Stanley Temple that stopped me in my tracks. It read: "De-extinction is essentially a game-changer for the conservation biology movement. It changes one of our principal arguments, that extinction is forever." Temple, a man I'd long regarded as a scientific authority, was all of a sudden saying that something I'd barely heard of was dismantling the least disputed tenet in my old favorite field. And that's when it really hit me. Undoing "forever"? De-extincting? Changing the direction that leads to the end? Had science finally found a way to make death reversal real? Spoiler alert: no. But there was something more subtle lurking out there, and I needed to find out what it was.

I MET WITH Stewart Brand on March 15, 2013, in a warm, well-lit room at the end of a labyrinthine set of corridors in the National Geographic Society's headquarters in Washington, DC. A white-haired fellow of somewhat lanky stature, Brand has a relaxed yet mighty presence. That day, Revive & Restore had organized a public TEDx speakers' event on de-extinction, bringing a megaphone to their mission for the first time. Talks about the science, ethics, law, policy, history, even photography and art, all wrapped up in de-extinction, were broadcast to screens around the world, and I was there to report on it.

Peering at me through wire-rimmed glasses, Brand told me how the idea for Revive & Restore first got off the ground. "My wife was acquainted with this guy George Church at Harvard," he says, explaining their connection through her career in biotech. George Church is a professor at Harvard Medical School and MIT, the head of the Personal Genome Project in the U.S.— which Phelan was peripherally involved with, and which aims to make human genomic data publicly available—as well as one of the world's most accomplished scientists in a variety of areas that touch on genetic engineering. I'm not just talking about an impressive CV: Church's team smashed the record for how many genes can be edited at once, when they altered sixty-two pig genes in one go. Church also figured out how to encode an HTML version of a book, one he'd cowritten, in a single drop of DNA, through digital biological conversion. And he's one of the leaders of the Genome Project-Write, an effort to assemble the entire collection of human DNA and that of other species in their own synthetic genomes. He is tall, vegan, and bearded—with a particular style of white facial hair that makes him look somewhat like a cross between Charles Darwin and God. Brand told me he learned through his wife that Church has techniques that allow him to take DNA from extinct animals "and basically swap it into the genomes of living animals, and turn the living animal into the extinct animal to get them back…" This, he thought, "has to be pursued, so I pursued it." The idea quickly took flight.

FOR THOUSANDS OF years, passenger pigeons lived in and flew across much of North America, from the east to the Midwest. They moved in massive flocks, billions of them at a time. It would sometimes take several hours for one flock to pass over any single spot. In a 1947 essay, Aldo Leopold writes of this species, "The pigeon was a biological storm… Yearly the feathered tempest roared up, down, and across the continent, sucking up the laden fruits of forest and prairie, burning them in a traveling blast of life." But by the end of the nineteenth century, the pigeons' numbers had dramatically dwindled. In less than fifty years, the bird that once turned the blue above black was gone. Then, on September 1, 1914, the very last passenger pigeon— Martha, a roughly twenty-nine-year-old female with a trembling palsy—died in the Cincinnati Zoo. She never in her life laid a fertile egg, but her story since her death has fertilized excitement for the idea that her gene pool might be recreated.

On May 24, 2011, Brand sent an inspired email to George Church, who he knew was interested in passenger pigeons, and to the great evolutionary biologist E.O. Wilson:

Dear Ed and George,

The death of the last Passenger Pigeon—in 1914—was an event that broke the public's heart and persuaded everyone that extinction is the core of humanity's relation with nature.

George, could we bring the bird back through genetic techniques? I recall chatting with Ed in front of a stuffed passenger pigeon at the Comparative Zoology Museum, and I know of other stuffed birds at the Smithsonian and in Toronto, presumably replete with the requisite genes. Surely it would be easier than reviving the woolly mammoth, which you have espoused.

The environmental and conservation movements have mired themselves in a tragic view of life. The return of the Passenger Pigeon could shake them out of it—and invite them to embrace prudent bio-technology as a Green tool—instead of menace in this century... I would gladly set up a nonprofit to fund the Passenger Pigeon revival...

Wild scheme. Could be fun. Could improve things. It could, as they say, advance the story.

And advance it, it did. Church allegedly got back to Brand in less than three hours, detailing how he believed they could return "a flock of millions to billions" of passenger pigeons to North American skies. Excited about the possibility, Brand and Phelan brought researchers from around the world who were interested in de-extinction under the auspices of their new nonprofit.

Since then, Revive & Restore has convened several meetings—large and small, public and private—at prestigious bastions of science like Harvard University's Wyss Institute for Biologically

Inspired Engineering and the National Geographic Society. They host a private online email discussion for those closest to the research and create educational materials for the public so that we can learn about what they're up to. They spend a lot of their time nourishing connections between experts who work on de-extinction and looking for research funds and donations to get projects they're interested in off the ground.

They insist that their goal with de-extinction is partially a democratic one: to engage us all in thinking about what the re-creation of extinct species might mean before any wind up in our national parks or zoos. They want those steering scientific projects to be able to adjust direction according to societal concerns and expert criticism. That kind of public engagement kicked off with the TEDxDeExtinction event and coincided with a cover article in the April 2013 issue of *National Geographic*. From that point on, the idea gained publicity at a faster and faster clip and was soon profiled by features in the *New York Times Magazine*, *The New Yorker*, and a flood of other publications. Much excitement and concern, hope and trepidation, support for and anger about de-extinction have been percolating ever since. And when I ask Brand how he expects to deal with the ensuing controversy, he replies, "We've got plenty of time to think about it, argue about it, regulate it." Time is on their side— luckily for them, because each de-extinction project is crawling with issues that require insightful planning.

I commend the gesture toward public engagement but wonder, is it anything more than that? What would public debate really have to look like in order to truly influence a possibly privately supported science? Does the nod to public dialogue establish real tools for society-led regulation, or does it merely tick a box that says, "The public has weighed in and therefore we can proceed"? More importantly, why should we be having a serious debate about all of this anyway? What is it that Revive & Restore really wants de-extinction to do?

The answer to the last question can be found in their mission: "to enhance biodiversity through the genetic rescue of endangered and extinct species." Traditionally, genetic rescue involves increasing gene flow into populations of animals that are suffering from low genetic diversity, done by directly inserting it through a variety of methods. Revive & Restore thinks of *genetic rescue* as an umbrella term that captures the multiple potentials of gene-editing, cloning, and selective-breeding techniques to restore impoverished ecosystems. Some of these techniques have more possibilities than others. And they aren't relevant only for extinct species, but for those that are still living as well. They call their genetic rescue work on extinct species *de-extinction,* and the work they're doing on endangered species *genetic assistance,* but their separation is much more taxonomical than philosophical. "I don't separate out our de-extinction work from our endangered species work or the work that we might potentially do one day around wildlife diseases to help endangered species," Phelan later tells me. That's because their take on genetic rescue is layered, like a cake: you've got to cut into it to see what it's made of.

The first layer is the determination of what is going on with a species at the level of its genes. By collecting and analyzing a species' genetic information, researchers may be able to identify what made a species vulnerable to extinction in the first place or what is making an endangered species particularly susceptible to a threat in its current environment.

The second layer is the editing of the species' genome to make it less defenseless against the dangers unveiled by genetic analysis. For example, virus-resistant genes could be inserted in the genome of a species that is dying from an infection, or genetic susceptibility to a disease could be altered with precision gene editing, which we'll explore.

The third layer is where de-extinction comes to life. Here,

as Revive & Restore's website says, "the trick will be to transfer the genes that define the extinct species into the genome of the related species, effectively converting it into a living version of the extinct creature." But to effectively convert a living thing into an extinct creature (in the opinion of Revive & Restore, at least), you don't need to make a carbon copy of the original. If it looks like a duck and acts like a duck, it's a good enough duck—or passenger pigeon, or woolly mammoth or heath hen or aurochs or great auk, as the case may be. When a species goes extinct, the role it once played in its ecosystem—as a fertilizer, forest disturber, predator, and so on—vanishes along with it. This absence leaves a niche unfilled, which in turn affects a cascade of other life forms that a species is always connected to, from the tiniest of mites to the tallest of trees. *Keystone species* are those known to crucially affect the overall function of an ecosystem, and they are generally the sort whose functions de-extinction advocates say we should reconstitute now by inserting into existing animals the critical traits that once allowed these species to live out their particular ecologically beneficial role. The emphasis in these proposals is on recognizing, reassembling, and inserting genes that code for those particular lost traits that may be desirable to have back, and not on recreating the exact original species, as if that were even possible.

It's been suggested that if the resulting animals are "good enough" replicas of the extinct species—largely judged by whether they can function the way the extinct species did in the wild—they should be reintroduced into habitats where they will restore the ecological roles lost with the species. Then the ecosystems will edge happily toward what they used to be like, and beneficial dynamics between the flora and fauna will return to fruition. *Maybe.* Or perhaps not. That idea may be built on a faulty premise, critics argue, because we still don't understand everything there is to know about how those past ecosystems functioned. When de-extinction advocates suggest that introducing a recreated species that can live and act like an extinct keystone species will restore an entire ecosystem to what it used to be like, many other crucial biotic factors risk being overlooked. The ecosystem function of a species is going to be dependent on learned behaviors that come from living with other individuals of its kind, not on morphology alone. This could be a problem when an unextinct species is born with the help of a bunch of human minds and laboratory tools instead of a natural herd that it can learn from. Plus, depending on how long ago the ecosystem changed, there are going to be dozens to hundreds of other species that would need to be reintroduced as well—from bacteria to big animals—in order to ensure that the ecosystem gets restored.

These issues can be picked apart in many ways, but first, a clarification. If taken without scrutiny, the term *de-extinction* as it is widely used suggests that reversing extinction might actually be achievable. That idea, however, is a sham. In no way can we ever undo the erasure of an entire way of life. Take the woolly mammoth, for example.

To create a woolly mammoth today, one that's identical to the one that traipsed across the Bering Land Bridge between Siberia and Alaska thousands of years ago, a genetically and behaviorally exact copy of the original woolly mammoth would first need to reappear. Artificially selecting or breeding the woolly mammoth's closest living relatives—which happen to be Asian elephants—to recreate their look and feel wouldn't work because elephants are too evolutionarily divergent from woolly mammoths and didn't directly descend from them. If we're talking about an *exactly identical* re-creation, editing mammoth genes into elephants is also out of the question: there'd be a slew of elephant genes in the mix. Cloning can get close, but it's not possible when tissue was not taken from the animal sometime before it died and immediately frozen. Woolly mammoths (and many other extinct creatures) have been gone far too long for any human to have intentionally preserved tissue with the foresight that they might be resurrected one day. And although some researchers are looking for perfectly preserved mammoth cells in order to clone them, no one has been able to find any yet. The technology scientists use now to clone living animals requires cells of the animal that are full and intact, with the DNA neatly packed away inside each cell's nucleus.

But "full and intact" does not describe the state dead organic things come in, especially when they've been exposed to the elements for hundreds or thousands of years. And crucially, the embryo that cloning produces needs to develop somewhere, but when all the females of a species are gone, the embryo must be embedded inside the womb of a surrogate mother from another species or subspecies, or perhaps someday in an artificial womb. All of these options are likely to introduce the developing fetus to a different set of hormonal and microbial interactions with the carrying mother (or machine) than those newborns of the extinct species experienced when they grew inside a mother of their own exact kind. There's also the issue of mismatching mitochondrial DNA, which I'll discuss later. My point is, any way you look at it, identical development is just not in the cards.

It is difficult to conceive of a situation in which an exact genetic and behavioral replica of a vanished species could ever reappear. But I think that what *is* possible is even more interesting than getting the identical original back. Scientists are learning how to cobble important elements of an extinct species into a new life using a few different methods. Various degrees of resemblance to the extinct creature might be achievable, depending on which method or methods are applied. But whatever the choice for how it's done, the result is a new organism—perhaps even a new species—that can live and act, with varying success, like an extinct species.

Some believe that the human-caused extinction of a species carries a moral imperative that bolsters the case for de-extinction. Brand has said that we should de-extinct species for the same reason that we protect endangered ones: to undo harm that humans have caused. The idea can be traced to this meditation of Gary Snyder's on a Zen Buddhist verse: "The precept against taking life, against causing harm, doesn't stop in the negative. It is urging us to give life, to undo harm." As Revive & Restore sees it, this rehabilitative tenet has been woven into the ideological fabric of de-extinction. "Humans have made a huge hole in nature over the last ten thousand years. Now we have the ability to repair some of the damage," Brand said in a 2013 TED talk. "Part of 'do no harm' is 'undo harm'... Want to try it?" Phelan later nuances her husband's words for me: "We are not motivated by guilt, which often goes along with undoing harm, but we are committed to trying to make the world a better place."

A better place? When I tell people I'm writing this book, they often do not see it that way. "We've heard this story before," they'll say, "and the ending to *Jurassic Park* was not pretty."

Jurassic Park, Michael Crichton's bestselling novel about dinosaur resurrection gone wrong, inspired a series of blockbuster hit movies that have instilled in the public consciousness ideas about the good and evil of human hubris. The moral of *Jurassic Park*—that chaos will prevail if humans act with self-conceit—played into a story we like to tell about how science has ramifications that can't be seen from the outset. But Brand has a swift response for people who are quick to compare *Jurassic Park* with what he and his colleagues are up to. Mainly, it's that they're a nonprofit and, according to him, not interested in hiding anything. "What drove the plot of *Jurassic Park,*" he tells me, "is all this private corporate secret. Maintaining the secrecy of the project is what let it become pathological."

Beyond the fictional fantasy, what is it about de-extinction that upsets people? To my mind, the potential for recreated animals to become invasive species once they're out in the wild, the chances of their hybridizing with extant species in ways that could jeopardize the unique genetic properties of those populations, the possibility of patenting and commodifying life forms, welfare issues for the experimental animals—both the ones being created and the ones already living that are used to make de-extinction work—and the idea that de-extinction might make people feel less concerned about species going extinct if it can bring them back at a better time raise serious concerns. I'll get into all of those and more. But something else has been nipping at my heels while I've been writing this book, something that I still can't fully reconcile.

Familiarity risks breeding fondness, and as people hear more about de-extinction, they may increasingly warm up to the idea that it might work. In that sense, that I am even writing this book is an ethical consideration. Am I merely adding one more narrative to the pile that will get people accustomed to de-extinction when they might otherwise be more opposed to it? Some of the people I interviewed for this book expressed disappointment in me for this very reason. Phelan rightly says that as their de-extinction outreach continues and the technology evolves, pressure will be relieved from the need to understand everything about this uncertain "new scary science." She expects that over time, people will respond with a shrug when they once might have shuddered, will adopt an attitude of "Well, let's see what they can do." I believe that, in all cases, de-extinction requires careful analysis of the pros and cons attached to each project's goals, and in some cases, there are aspects of de-extinction I can get behind, and am even excited about. But I make no blanket statements of approval here. Whether the candidate in question is a great auk (a black and white flightless bird with a heavy, hooked beak), a gastric-brooding frog (which gave birth to babies out of its mouth), a thylacine (a carnivorous marsupial with striking tiger-like stripes on its backside), a quagga (a coffee-colored relative of the zebra with a non-striped bum), an Irish elk (one of the largest deer to ever live), a woolly mammoth (the iconic proboscidean beast), a passenger pigeon (the bird that once flocked in the billions), an aurochs (the ancestor to all of today's domestic cattle), or something else, each species has a variety of qualities and requirements that require careful scrutiny from experts and the public alike. That's why I have written this book: to help people analyze the idea for themselves. But merely by putting a book about de-extinction out there, am I aiding and abetting a movement that I do not wholeheartedly support? Some people have told me that's exactly what I'm doing—it's a game of cat's cradle that I'm still trying to sort out.

Other books have been written about this topic since Revive & Restore was founded. The first was by Beth Shapiro, a leading scientist in the field of ancient DNA research and the codirector of the Paleogenomics Lab at the University of California, Santa Cruz. Her book is called *How to Clone a Mammoth,* but ironically, it concentrates on how we can't. When she was asked why she would call the book something so misleading, she said, "I probably should have called the book *How One Might Go About Cloning a Mammoth (Should It Become Technically Possible, and If It Were, in Fact, a Good Idea, Which It's Probably Not),* but that was a much less compelling title."

You might have noticed that the word *de-extinction* doesn't exactly roll off the tongue. In her book, Shapiro prefers that we call the end product of a de-extinction process an *unextinct species* rather than a *de-extinct* or *de-extincted* one. None of these terms are ideal, but the notion of an "unextinct" species implies that parts of a long-lost group have been revived, whereas a "de-extinct" or "de-extincted" species implies that the overall process of extinction has been reversed. From here on in, to create some consistency, I will use the term *unextinct* (rather than *de-extinct*) to describe an animal or species that has resulted from the technologies and plans that are put forth in these pages and *de-extinction* to refer to the overarching movement, its practices, and its aims.

Shapiro thinks there is much more public fear and resentment about the researchers involved in de-extinction than there would be if people understood how fringe their activities really are. In an attempt to clarify the scale of their work, she once said in an interview, "I think there's a huge misconception about how much science is actually going on. In the back corner of George Church's lab [at Harvard] they have a few people who are using a tiny amount of resources that are available to them to attempt to swap out genes in elephant cells which are growing in culture in a dish in a lab. I have a student who's trying to convince me that it's a good idea to bring passenger pigeons back to life. There's a group in Australia who are thinking about the gastric-brooding frog but are stuck because they can't cause the cells to actually grow up. There's a group in New Zealand that is thinking about bringing a Moa [an extinct bird] back to life and are working on sequencing the moa genome, which is not de-extinction, in itself. There's a Spanish group that's thinking about the bucardo [a subspecies of Spanish ibex that went extinct in 2000], and there's the backbreeding group for the aurochs [an extinct species of wild cattle] in Holland. That's it. That's everything that's going on in the world right now." And at the time that I'm writing this, that's still pretty much the case, with a few exceptions. But it's not the abundance of research in this area that's so striking—it's how well above its weight it punches in the fascinating issues it creates.

That's partly why this book picks up where a discussion focused on its technical basis would leave off. Although I am captivated by the science and technologies that make this movement possible and do explore them in the pages that follow, my larger focus is on the cultural, ethical, environmental, legal, social, and philosophical issues that de-extinction sets free into our world.

I begin with an overview of the methods used in de-extinction, before exploring topics such as human-caused extinction, the choice of candidate species, de-extinction projects currently underway, the rewilding of ecosystems with recreated animals, protection of reintroduced species, patents on and profits from de-extinction, benefits for endangered species, lessons from species that are already gone, and personal stories from revivalist researchers. To that end, I've spent the last few years exploring the labs, lore, and laments of a group of scientists who want to nuance the tenet that "extinction is forever" as well as to help fauna on the brink. This book is a result of uncountable discussions I've been fortunate to have with some of those leading the science, as well as with philosophers, ethicists, artists, historians, legal experts, critics, and opposing scientists. It is about the humor and the hope, the application and the absurdity, the elegance and the empiricism that tie this movement together. And it is about why this wild idea may or may not have more meat on its bones than *Jurassic Park*'s dinosaurs ever did.

CHAPTER ONE

HOW IS DE-EXTINCTION DONE?

Life will find a way.

—MICHAEL CRICHTON, *Jurassic Park*

Deep Inside Ancient Amber

IN 1983 IN Bozeman, Montana, a cluster of genetic researchers gathered for the inaugural conference of the Extinct DNA Study Group. The purpose of the conference was to discuss the recovery and deciphering of DNA from extinct organisms. Among those attending was George Poinar Jr., an entomologist and parasitologist with a particular fondness for the prehistoric creatures which later evolved into birds that fed on the insects he studies. He and the others discussed new possibilities for preserving ancient organic materials from long-dead life forms. It was your typical academic scene. But before the conference was over, they'd come up with a concept that they had no idea would end up driving Hollywood hits.

At home, Poinar had an impressive collection of insects that sat ever so still in hardened amber blobs dating back thousands of years. His son Hendrik remembers that when he was a child, his father would sometimes pull them out at night to examine them under the light in his study. Hendrik would watch his dad rub the amber with oil and inspect its contents under the microscope. He'd slink under his father's arm and crawl up into his lap to get a closer look, squinting as he tried to get the oculars in just the right position over his eyes so that he could make out the bugs in 3-D. "There would be spiders and flies that looked like they were not a whole lot older than a few minutes," he says, now fully grown, with a lab and family of his own. "It looked like they'd just been caught flying or eating, their cilia still standing upright, and wings with full venation." It seemed as though an insect could come to life at any moment, twitch to release itself from the golden goop, and fly away.

Back in Bozeman in 1983, one of the researchers mentioned that George Poinar Jr. had dated some of his amber to the Cretaceous and that many of the most interesting pieces may have encased petrified ancient mosquitos. If you were to find some of the bloodsuckers that had been caught in the act, indicated by their engorged red bellies, the scientists reckoned, those particular Cretaceous mosquitoes might have lived at the same time as the duck-billed dinosaurs. That got them thinking about what the relationship between those species might have been. The mosquitos' bellies could only have been filled with blood, and now they had an idea about whose blood that might be.

Because amber is anhydrous—a fancy word for "very dry"—and because dry environments are ideal for preserving organic materials, the researchers speculated that some of the mosquitos' cells might still be preserved inside the amber nuggets. If that were true, it was plausible that some of the mosquitos' DNA could still be inside those cells. And if the DNA of a mosquito could be preserved in amber, then logic would say that the DNA of the dinosaur the mosquito fed from might still be in its abdomen too. This idea was spurred by an article in *Science* in which Poinar and his wife, Roberta Poinar, showed that they had found organelles—ones that normally contain DNA—still intact inside of a 40-million-year-old fly preserved in Baltic amber.

The scientists in Bozeman further speculated that if that fly did indeed contain dinosaur DNA, it might be possible to recover it from a dinosaur blood cell removed from the belly of the resin-encased insect, making it possible to reconstruct the genome of that dinosaur by reverse-engineering the genetic code of the recovered DNA strands. If one could extract the DNA from the dinosaur blood cells trapped inside the bug trapped inside the amber and then read out the long string of molecular bases that constitute DNA—adenine, cytosine, thymine, and guanine, or A, C, T, and G for short—it might be possible to resurrect the creature from its code. Some steps about repairing damaged DNA with genetic material from another living animal were glossed over; the exact protocol was murky in the researchers' minds. But they were more interested in having fun with their idea than proving the theory, so they wrote it up and published it in a small academic newsletter. It was just a far-out concept about a futuristic technique, not one they expected people to take all that seriously. But one reader thought it was seriously entertaining and had some ideas to add.

Somehow, a writer by the name of Michael Crichton (you may have heard of him) got his hands on the paper and contacted Poinar, asking if he could pick his brain for a book he was writing. Poinar willingly agreed, but Hendrik remembers that his father kicked himself later, wishing he'd seen the bestseller potential himself—early retirement would have been nice, after all. But he was humble enough to know that Crichton did a much better job than he could have at bringing this story out of the lab and into the public's imagination.

Resurrecting a dinosaur from ancient DNA in amber might have worked for fiction and Hollywood, but most of the science says that it won't work in the real world. In the 1990s, some researchers claimed to have extracted insect DNA from amber that was roughly 130 million years old. That caused a huge stir in the ancient DNA community; others tried to repeat the experiment, but those results were never found again. In cases where scientists have claimed to have recovered DNA from ancient amber, it is now largely thought that their samples were contaminated with modern DNA that could have come from the hands of the researchers themselves or even the sandwiches that they ate for lunch. However, George Poinar Jr. tells me that "every amber sample is different in regards to its potential for preservation. If the insect was exposed to the sun while stuck to the surface of the resin before it was eventually completely entombed, then it would have a poor chance for its DNA to be preserved. If the sample sat around on the desktop for months, then small amounts of atmospheric air could have entered and degraded the remaining DNA. Since no two amber samples are the same, nobody can definitely say that the original results were erroneous, just that their results were unsuccessful, perhaps by using sloppy technique."

It has been widely stated by scientists that it is impossible to salvage DNA from ancient amber (the studies that happened in the 1990s were falsified). In her book *How to Clone a Mammoth*, Beth Shapiro goes to great lengths to explain why, as she takes her readers through an experiment of her own. But Hendrik Poinar, a respected paleogeneticist himself, has other ideas. He says, "I haven't given up complete hope on amber. I think the place to start is really to look at copal, which is very young amber, and look within the timeframe where we think DNA molecules are still preserved—so around 10,000 or 30,000 years old." Although Shapiro has also tested copal, Poinar remains curious about it. "I don't think the door is closed on amber despite what people say." His father agrees. Perhaps, one day, one of them will reopen it.

If you've read *Jurassic Park* or seen the films, you'll know that the amusement park's mastermind, John Hammond, was only able to fill the place with dinosaurs thanks to a formula that sounds a lot like the one that George Poinar Jr. and his colleagues came up with and that his creator found in a newsletter. In the story, Hammond has access to amber nuggets that encase bloodsucking insects from the time of *T. rex,* which allow him to reverse engineer his own dinosaurs. But in real life the dinosaur would have to be built back up from its exact genetic blueprint. Its DNA would need to be whole and intact, with the right genes lined up in the right order, and be able to be read out as a complete genomic map. If you think of a genome—the total collection of genetic material in an organism—as a big puzzle, you would need to have every single piece of the puzzle and each one would need to be put in the right place. But DNA is an organic material that decomposes like any other: pieces of the puzzle disintegrate over time.

In fact even at the best of times, ancient DNA comes nowhere close to the stability of a puzzle piece. DNA starts to degrade as soon as a cell's oxygen levels start changing. When an animal dies, the cells go through autolysis, in which the pH balance that regulates enzymes inside the cells gets knocked off kilter. The moment a creature stops breathing, its oxygen concentration changes, the pH in the cell shifts, and the enzymes go bonkers, chopping up the DNA inside. Unless you flash-freeze the cells in liquid nitrogen immediately upon death, that cellular chaos cannot be avoided. One study has shown that it is possible to clone mice from mice brain cells that had been frozen for sixteen years without any substance to protect their cells from the damaging effects of –20 degree Celsius temperatures (–4 degrees

Fahrenheit). Another study demonstrated that a bull could be cloned using cells from bull testicles that had been cut off after the animal died and that were frozen at -80 degrees Celsius (-112 degrees Fahrenheit) for ten years. The bull the testicles came from was named Yasufuku, a prized sire whose offspring grew into beautiful bovine specimens that could be farmed for their marbled Wagyu beef. Twelve hours after Yasufuku died of old age, his testicles were collected, wrapped in aluminum foil, placed in a freezer and also weren't treated with any protectant substance. But mice—and disembodied testicles—are relatively small and can freeze rapidly. Big prehistoric beasts would take much longer to freeze solid. The dinosaurs didn't live in an icy world anyway; they've been preserved as fossils in rock. To paraphrase Robert Lanza, the first scientist to clone an endangered species, "You can't clone from stone," so you won't find anyone suggesting dinosaurs as realistic de-extinction candidates.

The integrity of DNA also gets lost during fossilization. Chemical changes in its makeup create sequence errors, and its strands break, creating holes in the code, which then can't be read thousands of years later. Animal fossils, teeth, and bones can be porous and so may soak up water teeming with microbes. The DNA inside breaks down over time and becomes contaminated with the DNA of other organisms in its midst, like bacteria and fungi. Sometimes the contamination is so rich that paleogeneticists are lucky if even 1 percent of the DNA pulled out of a fossil is from the extinct organism they are interested in. So unless the puzzle pieces you play with look like soggy papier-mâché, I'll admit that this metaphor is a bit of a stretch. Hendrik Poinar has a better one. He should know; he followed in his father's footsteps to become a scientist and now runs his own ancient DNA lab at McMaster University, in my home province of Ontario, Canada. He tells me that dealing with ancient DNA "is like taking the entire collection of *Encyclopedia Britannica*, ripping it up into two-letter pieces, scrambling it all up, and then having some grad student put it back together without coffee."

Jurassic Park is a gripping tale, but neither the book nor the movies have influenced the science. Depending on how you see it, the reality is either far scarier or more banal. Scarier because de-extinction is becoming feasible, which means humans will be solely responsible for the well-being of the experimental animals that we recreate as well as for the precious environments they're introduced into; more banal because it might all end up going according to plan. But no matter which way that falls out, no dinosaurs are on the list for possible de-extinction, and they never will be. For a species that has vanished relatively recently, museum specimens may do the trick. And in cases where exceptionally well preserved remains of extinct animals have been chilling for thousands of years, like frozen woolly mammoths in the Siberian permafrost, chunks of DNA can sometimes be salvaged. But the dinosaurs did not fall into ice when they died, and they've been gone far longer. They vanished approximately 66 million years ago—so long ago that it's too challenging to work with their DNA. The oldest ancient DNA that has been salvaged to date came from a 700,000-year-old horse, which is roughly 1 percent as old as the most recently living non-avian dinosaurs. The science here is in its earliest stages, and even that's not quite like it is in the movies.

What Is a Species?

IN ANCIENT GREECE, the word *species* applied to objects as well as to living things: it referred to different kinds of *things* that share similar physical properties. Taxonomists and philosophers did not even make a distinction between inanimate objects and biological beings. Species were classes of categories, separated by their different characteristics, and any abnormal individuals were dismissed as mistakes. Until the nineteenth century, the most important criterion for identifying a species was simply what it looked like. That thinking, known as the *morphological* or *typological species concept,* prevailed until it became clear that physical differences alone could not always cut the mustard. Researchers discovered many creatures that were hardly different in their appearance but that had many genetic differences between them, making species identification cryptic and confusing. Those beings that were undetectably different to the eye but genetically distinct in their cells became known as *sibling species.* Sibling species could be as *genetically* unique as species assorted by the morphological species concept, but they could look the same. In 1975, a biologist named Tracy Sonneborn discovered fourteen sibling species among individuals of a protozoan (a type of single-celled organism), *Paramecium aurelia,* that was considered a single species according to the morphological species concept. A single species could also have several different body types, especially at different stages of maturation—a butterfly is one example. So what was the one factor that defined a species? That thing was sex.

In 1942, biologist Ernst Mayr proposed the *biological species concept.* Although this theory had been foreshadowed by other thinkers, the concept, which still dominates much of today's species science, is largely attributed to him. According to Mayr's definition, a species is the largest group of organisms in which two individuals are capable of reproducing fertile offspring. If two animals can successfully mate, creating a new generation of animals that can reproduce themselves, then they must belong to the same species. If two animals of different species mate, various factors can prevent successful fertilization or successful birth. In some cases, the offspring might be sterile, as when a donkey and a mare mate and produce a mule, which cannot reproduce. If nature wanted to keep its inventory neat and tidy, the biological species concept would be its perfect tool—a tool too good to be true.

No species concept has been universally applicable across all living taxa. Depending on what concept is used, scientists end up with an entirely different list each time the count is made. Species don't always stick to their own kind when choosing a mate, and unlike the mule, some hybrids can have offspring of their own. A bat from the Lesser Antilles is believed to have evolved from three distinct bat species, for example, one of which is already extinct. Similarly, the small Clymene dolphin arose after two other Atlantic dolphins of different species—the spinner dolphin and the striped dolphin—rubbed fins. If you drill deep into our DNA, you can even see that many of our ancestors successfully mated with Neanderthals.

In 2010, a group of researchers at the Max Planck Institute for Evolutionary Anthropology in Leipzig made headlines when they discovered that humans and Neanderthals had interbred, and published their discoveries in a paper with Richard (Ed) Green as lead author. Further research, however, showed that our cross-species coitus was rare, and in some cases, maladaptive. It's believed that some of the genes we acquired from Neanderthals might have made us susceptible to diseases like lupus, Crohn's, and type 2 diabetes, and when we did interbreed, our offspring could have had fertility problems. But our interbreeding may have also benefited us in the face of environmental change, boosting our immunity against bacteria, fungi, and parasites.

Approximately one-fifth of the Neanderthal genome is said to be detectable and dispersed throughout the genomes of non-African humans today. The personal DNA analysis company 23andMe can even tell you how much Neanderthal DNA you carry in your genes. All you have to do is buy a package that they'll send to your house that contains a tube you're supposed to spit into. Do the deed and then mail it back. An acquaintance of mine was inspired by this service to dream up a dating app for people with above-average Neanderthal DNA. His idea? They meet, they have sex, and after several generations, Neanderthal-like babies will materialize from the density of their hybridized genes.

George Church takes the prospect of Neanderthal de-extinction even further in his book, *Regenesis*: "If society becomes comfortable with cloning and sees value in true human diversity, then the whole Neanderthal creature itself could be cloned by a surrogate mother chimp—or by an extremely adventurous female human." His comments ignited a swarm of controversy, but Church insists that he was taken too seriously and that he knows societal acceptance must come before any enthusiast's wish to volunteer their uterus.

We still tend to speak of Neanderthals as though they're somehow alien from us, even though many of our genomes show that they're literally a part of us. To be a species is a murky thing. Many of us are mutts already, but de-extinction could multiply the spectrum of hybridity in some startling ways.

How Are Scientists Trying to Make Unextinct Species?

BEFORE WE SOAR along the spectrum of de-extinction research happening around the world today, here is a brief explanation of the specific technologies that are making it possible. Some of the technology does not need much explanation, however, because we've already been using it for centuries.

SELECTIVE BREEDING

In a debate about genetically modified organisms, it is not unusual to hear an advocate say something like "It's really no different from what we've been doing for hundreds of years. Think about cheesemakers or brewers, for instance. They've been manipulating life forms for eons, coercing all those bacteria and yeast to make things for us to eat and drink. Or no, even better—think of dogs! Where did they come from? We intentionally bred them from wolves, turning them into everything from Great Danes to Chihuahuas for our own liking. How natural is that? You don't seem to have a problem with your dog when you take it out for a walk." The bark of this continuity argument, which points to breeding practices, can be heard wherever genetic modification advocates are on the defense. Supporters of genetic modification place it on an age-old continuum to make the point that humans have a deep history of sculpting life into the shapes we like. It then follows that our continuation of that practice now should not be seen as a new threat. Since only the tools, not the purpose, are novel, any potential danger is dismissed. Familiarity is used as a welcome mat to invite skeptics into the fold.

The domestication of dogs involved choosing wolves with the characteristics we liked, then mating those wolves with each other to create new animals, sharpening those desired characteristics in each new generation. This type of human intervention is called *artificial selection* or *selective breeding*. With canines it became so sophisticated that one day Crufts— the world's biggest dog show—bizarrely appeared. And yes, as those advocates note, domestication through selective breeding or artificial selection has a long history. Charles Darwin's *Variation of Animals and Plants under Domestication,* published in 1868, demonstrated our creativity in cultivating an organism's physical traits. By selecting a trait based on appearance and then mating animals or plants that have the best representation of that trait, breeders became nature's curators.

Take the quagga, for example, a close relative of the zebra that lived on the plains of South Africa. It had dark stripes just like a zebra's all over its body, except on its backside, which was solidly mocha-colored. Its stripes lay on top of a light coffee coat and so lacked the stark contrast of the zebra, with its iconic white underlayer. But quaggas were often seen with zebras in the wild and were often confused with them—so much so that when the last quagga mare died at the Amsterdam Zoo in the 1880s, no one realized that her death marked the extinction of quaggas altogether. Zebras were still roaming around the African plains, and it took a while before people realized that the quaggas had been a related group of animals that were ruthlessly hunted until none were left.

Today, we know that the most northern subspecies of the plains zebra is the striped black and white beauty you are familiar with and that the southernmost subspecies of the plains zebra was this tawny, half-striped quagga that never saw the dawn of the twentieth century. But at the time of the quagga's extinction, no one was sure if it was a subspecies of zebra or a different species altogether. There were no records indicating whether zebras and quaggas ever successfully mated, so zoologists searched out morphological evidence to try to understand how the animals were related to each other. They compared the cranial structures of quagga skeletons that had been kept in museums with the cranial structures of other zebras, and at first, the results suggested that they were different enough to qualify as separate species. Then, a later study found that quagga crania and the skulls of zebras that lived in their range were very much alike. The museum specimens—the best quagga remains there were to work with—gave conflicting results. But the investigation took a turn when German natural historian Reinhold Rau got behind the wheel.

In 1959, Rau was hired to work at the South African Museum in Cape Town as a taxidermist and was excited to find that the museum had in its collection one of the world's twenty-three

RAU QUAGGA

remaining preserved quagga skins—the only one on the African continent. However, as he explains in his memoir, "It had been crudely stuffed in 1859 and appeared in a sad state... It was simply 'stored' in-between many other stuffed specimens, and even largely hidden by them. This worried me and the idea of improving its appearance and the way it was displayed came to mind." The stuffed quagga was a foal with all the wrong proportions, which he wanted to rightfully restore. Wetting the skin to do so would run the risk of destroying it, though, so he practiced on salted dry skins of a jaguar and some zebras until he eventually gained the confidence to dismantle the quagga foal and put it back together again in a more lifelike pose.

In the process, he discovered that some soft tissue was still clinging to the inside of its skin; he took it out, together with the

skull and foot bones. Even back then he had a feeling that something could be learned about evolutionary quagga relationships
from studying the soft tissue, and he tried—unsuccessfully—
for several years after that to find capable scientists who were
interested in working with the soft tissue he had. Along the way,
he kept track of his rejections, like the one from noted English
zoologist W.F.A. Ansel, who told him in 1979, "Unfortunately,
as I suspected would be the case, there does not seem to be any
meaningful cytological tests that could be carried out on such
long dead material."

In 1971 Rau traveled to Europe to visit most of the museums
there that housed preserved quaggas. As he examined, photo-
graphed, and measured each of the specimens he was shown,
he thought about their close similarity to southern plains zebra
populations he had seen in Etosha, Namibia, and Zululand. The
animals' strong resemblance to each other gave him the feeling
that the quagga might be a variation of the zebra rather than a
totally distinct species, and some of the people he met on that
trip encouraged him to investigate that idea further.

In the early 1980s, he made contact with interested genetic
researchers, including Russell Higuchi at the University of California, Berkeley. Higuchi and his colleagues took a sample from
what Rau had cut off his mounted foal and tried to extract the
DNA. By that time it was understood that the genetic relationship between two species could be determined by comparing
DNA sequences from an individual of each group: the more identical DNA the individuals had, the more recently they must have
diverged from their shared common ancestor, and therefore the
more closely related they must be. This technique had never
been used before with an extinct species, since it was believed
that the DNA would be far too degraded to make any sense of.
But in a scientific first, Higuchi and his colleagues managed to
sequence DNA from the skin of the extinct quagga mount. As

George Poinar Jr. writes in a 1999 article in *American Scientist*, the results "struck like a bombshell": ancient DNA could be analyzed, and eventually maybe even be cloned. That discovery launched the field of ancient DNA research, known as *paleogenetics*, without which there would be no de-extinction today.

Higuchi and colleagues went on to compare the DNA inside the quagga cell's mitochondria—organelles that serve as little energy factories—with the DNA inside the mitochondria of the plains zebra. The results were more similar than some people expected. All of the molecular evidence suggested that the quagga was not a different species from the plains zebra but a subspecies of it. This was no minor finding. It not only meant that Rau's hunch was right but also gave him reason to think that a radical idea he was considering might actually work.

When he was a schoolboy in Germany, Rau saw an impressive whitish bull with dark dapplings on display at a circus. That day he learned of the experiments of two brothers, Lutz and Heinz Heck, who were developing breeding techniques to create animals with ancestral traits that had disappeared as they'd become domesticated. Before the outbreak of World War II, Lutz and Heinz ran the Berlin and Munich zoos. There they selected individuals of various cattle breeds that they perceived to be of similar appearance (in color, shape, and horn size) to the extinct ancestor of domestic cattle, the aurochs. They then mated these individuals to try to unite their physical characteristics in new generations of cattle that resembled the aurochs as much as possible. But because they were trying to make prototypical German animals of a "strong and pure" nature, the cattle that the Heck brothers produced have been nicknamed "Nazi super-cows."

The brothers also tried selectively breeding other animals this way to resemble their extinct forms. Noticing that some of the plains zebras had reddish-brown hindquarters with fewer stripes than the average black and white zebra, Lutz had once suggested that selective breeding of plains zebras with the most quagga-like coloring and striping might eventually create animals that look a lot like quaggas. Rau was inspired by this, and, with the encouragement of other quagga experts he met on his trip to Europe, decided to pursue the idea. It's an idea he really went somewhere with. And since 1987, an initiative called the Quagga Project has been following in his footsteps on the path to breeding the quaggas back.

The Quagga Project's focus is on how the animals look rather than on their genetic purity. And although the right brownish color is hoped for, the correct striping has been deemed more important. By selectively breeding individuals from a "founder population" of the first chosen plains zebras with well-striped fronts and solid backsides and then mating them over the last three decades, researchers have begun to get some animals that look quite a lot like the original quaggas. Rau is no longer alive to see the results, but animals have been born with the right appearance and ability to live in the quagga's native range, and when that happens, they are given his name. But it is worth noting that the breeders of today's "Rau quaggas" do not claim that they've recreated the original quagga. Instead, they say they have created a whole new animal in its image.

Sometimes, when a population lives across a very large area, as quaggas and zebras did, individuals can become geographically separated, and differences start to grow between the dislodged groups. If a physical barrier like a mountain range separates one part of the population from the rest, the fragmented groups can, over time, evolve into separate species. But in some cases, even when two parts of a population are separated from each other, they are still—like the quagga and zebra—capable of producing offspring together. Groups that may have evolved to look different but are still able to reproduce with each other are considered subspecies of the same species rather than formally

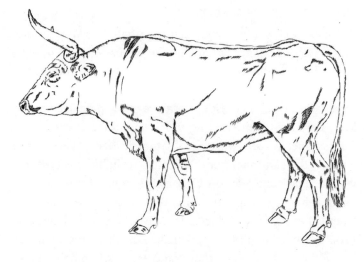

AUROCHS

distinct species. Therefore, even though the quagga and the plains zebra look different, they could probably still (if the quagga was around) get it on and produce offspring. The genetic tests that were performed on the museum skins were critical to making the Rau quagga possible: if the quagga hadn't been a subspecies of zebra but a distinct species all of its own, then the genes that code for the quagga's unique traits probably wouldn't be around in animals that can be bred today.

But genes that code for unique traits can sometimes persist after a species is gone—not only in their related subspecies but also in the gene pools of any direct descendants it may have left behind. The specific process of selectively breeding domestic animals in an attempt to recreate the traits of their wild extinct ancestor is known as *backbreeding* or *breeding back,* and that's what is currently being done in the name of recreating the aurochs.

BREEDING BACK THE AUROCHS

"My aunt was totally demolished by a bull. She was seventy-nine," Henri Kerkdijk-Otten—tall, brown-haired, and friendly Dutch historian by training and software test coordinator by trade—tells me one night over Skype after putting his children to bed. When he's not testing software or hanging out with his kids, Kerkdijk-Otten is trying to recreate the aurochs, which went extinct in 1627, when the last female died in Poland, and which is the early ancestor of all cattle, including the type that wounded his aunt. Ever since her awful encounter, he has been amazed by the strength of cattle. Whenever he's walking with a herd of cows, they tend to obediently follow behind, but if he turns around, they all scamper off like skittish kittens. That they should be so timid when they could easily kill him fascinates him most of all.

You can read about the aurochs in texts that stretch back to Roman times. But the best place to get a sense of what the originals were like is in the Paleolithic cave paintings of the slender beasts, with their S-shaped back and sloping hips. The calves were born red and transformed into black bulls or reddish-brown cows as they grew. Their remarkable horns could grow up to 43 inches long on the males, while the females' were just over half that. And a bull could become as tall as 73 inches at the shoulder—roughly the size of today's European bison, a species that lives in part of the aurochs' former geographic range.

The aurochs is the wild ancestor of taurine and zebu cattle, the two groups into which all living cattle breeds fall. Taurine are commonly found living in temperate zones, while zebu do better in hotter, tropical conditions. Working with different breeds of both types, Kerkdijk-Otten is using similar selective breeding techniques to those of the Quagga Project, but they are different in one important respect. Since all living cattle, not just one subspecies as with the plains zebra, are the aurochs' direct descendants, many of the aurochs' genes are still around, scattered in chunks across the genomes of today's cattle lines. This means the aurochs should be able to be "bred back to life" by selectively mating together different living cattle species, all of which have some bits of the aurochs' genome inside of them—presumably particularly the ones that look the most like aurochs.

Kerkdijk-Otten is confident that the aurochs' fragmented gene pool exists in its entirety among the animals he's working with. "All pieces of the puzzle that we need are still present in all living cattle breeds; we just have to follow the trail back." Following the trail back, as he says, or backbreeding, is the attempt to recreate the extinct "wild type"—the typical form of a species as it occurred in nature—of a domesticated animal breed or breeds using selective breeding techniques.

A few years ago, Kerkdijk-Otten started an organization called True Nature Foundation, which aims to recreate productive ecosystems with animals that Europe has not seen for a very long time reintroduced. One of the group's goals is to entice tourists to visit these ecosystems in different parks across the continent. His foundation will need to work with an established network of nature conservancies like Natura 2000, which has established an array of protected wildlife areas across Europe. True Nature Foundation has already identified viable project locations in the Netherlands, Spain, Portugal, Romania, Ukraine, and Poland. The backbred aurochs is just one of several species that the True Nature Foundation is reintroducing into those parks; others include the wild horse, the European bison, and the water buffalo. The group's slogan is "From restoration to revenue," the idea being that ecological restoration projects will be combined with money-making ventures such as nature-based tourism, and, in the case of the aurochs, the sale of unextinct flesh as wilderness meat. The breeding program—called the Uruz Project—launched in 2015, and it is expected to take twenty years before the populations of backbred aurochs will be self-sustaining and require little to no management from the breeders. Besides that, Henri has also affiliated himself with another, newer aurochs backbreeding project: the Taurus project, operating in Germany, Denmark, Latvia, Hungary, and the Netherlands.

Before founding the True Nature Foundation, Kerkdijk-Otten was involved in yet another aurochs backbreeding initiative, the Tauros Programme. But he thought that its breeding strategy was not optimal, so he left to set up a new project that uses different breeding lines. Today, he's trying to make larger aurochs than any other breeding program has attempted, so that the animals won't be easily outcompeted by bison if they ever meet again in the wild.

In order to get the aurochs he wants, Kerkdijk-Otten and his team have selected multiple cattle types that have traits closely resembling those of the aurochs and have organized them into different breeding lines. The cattle types have been selected because, for example, they have the same degree of sexual difference between the males and females as the aurochs did, or for their impressive horns and correct coloration. Once the desired traits are stabilized in new cows that result from mating those cattle types, offspring from each line will be mated with each other to produce a new generation. After doing this many times and much trial and error, the researchers eventually hope to get an animal that appears—and acts—like an aurochs. Like the Rau quagga, however, a backbred aurochs will always be more artful than unalloyed. Eventually, he may use gene-editing techniques to try to make the backbred aurochs genome as close to the real thing as possible. But for now, he's putting first things first.

Each generation of new aurochs will take about three years to make a new generation, which includes a nine-month gestation period and roughly two years of breeding. Kerkdijk-Otten says they expect to see the birth of what will be considered their first revived aurochs somewhere around 2020. Initially, the animals will be bred on farmlands, and then the individuals that seem best suited for ecological restoration will be selected and transported to nature reserves. The rest of the herd of suboptimal aurochs will stay on the farm until they are slaughtered. About 90 percent of the males will be killed for their organic meat, which will be sold to buyers who want steaks from cows that have not been dehorned or fed antibiotics. When I ask Kerkdijk-Otten what his favorite way to eat an unextinct aurochs is, he gives it a minute of thought, then smiles and says firmly, "Pan-fried hamburger, slowly cooked in its own fat."

In addition to profiting from tourism and the ecological meat market, True Nature Foundation wants to sell aurochs hides, horns, and skulls as wall hangings, supply the niche experimental archaeology market with unextinct bones, and provide live animals to zoos and wildlife centers for purchase. Is the True Nature Foundation just business as usual in ecological disguise? Make no mistake, Kerkdijk-Otten tells me: their first priority is to renew a sense of wilderness across Europe that it now sorely lacks.

CLONING

Animals have three types of cells: somatic cells, germ cells, and stem cells. The first, *somatic cells*, are common specialized cells that make up most of the body, like skin, liver, or blood cells. Somatic cells are diploid, meaning that they carry two copies of each chromosome, one from the organism's mother and one from its father. The second, *germ cells*, are what give rise to gametes—sperm and eggs—in sexually reproducing organisms. Germ cells are haploid, meaning they have only one copy of each chromosome; when a haploid sperm and a haploid egg combine, they produce a diploid zygote, which later develops into an embryo. Importantly, a zygote contains some cells that are completely undifferentiated, called *stem cells*. Where *differentiation* refers to the process whereby cells, tissues, and organs gain special features during development, *undifferentiated* cells are those that lack special features. Some stem cells can turn into absolutely any cell type of the body as well as placental cells; these are called *totipotent* stem cells. As the zygote develops into an embryo, those cells lose their ability to become placental cells but are still able to become any type of cell in the body and are called *pluripotent* stem cells.

So as we walk through the science of cloning, please keep in mind these three main types of cells: somatic cells, which are specialized cells of a certain cell type, like blood or muscle cells; germ cells, which give rise to sperm or eggs for reproduction; and stem cells, which have not yet been specialized and have the potential to become specific types of cells in the body.

Cloning allows for the creation of nearly genetically identical copies of an organism. It is done by taking the nucleus—an organelle that contains most of a cell's genetic material—from a somatic cell of the organism being cloned and transferring it to an unfertilized egg cell from the host: another organism's egg that's had its own native nucleus removed. Once the somatic nucleus is fused into the egg cell, it gets reprogrammed by its host egg cell. This must happen because the somatic nucleus came from a cell that was highly specialized to express the genes that allowed that cell to be a specific type of functional cell, in skin or muscle from the animal it came from, for example. And that nucleus must be reprogrammed back to what it was like when it was once part of an undifferentiated cell type—an embryonic stem cell. Incredibly, the egg cell is able to reprogram the somatic nucleus by erasing and remodeling the structures that made the somatic nucleus specialized. Once the reprogramming is complete, the egg cell is stimulated by an electric shock to start dividing. The now-viable dividing egg cell is implanted inside the womb of a surrogate mother, where, if all goes right, it will grow into a clone. This procedure, called *somatic cell nuclear transfer* (or SCNT), is the same process that was used to make the world's most famous adult animal clone: Dolly the sheep.

Dolly was created with SCNT at the Roslin Institute in Scotland in 1996. But Ian Wilmut, a British embryologist, had to modify 277 eggs this way before Dolly was successfully brought to life. The nuclear DNA used to clone Dolly came from a cell in a six-year-old Finn Dorset sheep's udder, and the use of this white sheep's "breast" material is how Dolly got her name ("Clone-ene, Clone-ene, Clone-ene, Clone-eeeeene, I'm begging of you please don't take my man"). This technique, used on those 277 eggs, resulted in 29 embryos, which were transferred into 13 surrogate mothers. In the end only one cloned sheep, Dolly, was created.

The fact that it took so many attempts to achieve success is not unusual with cloning, and that raises ethical questions about cloning extinct animals. Because of complications in development, lots of embryos are sacrificed in the cloning process. Many that grow into fetuses die soon after birth. Congenital diseases are common in clones, so they may suffer fatal complications before they leave the lab. Nobody knows how many lives or potential lives will be lost in the attempt to clone extinct species until these techniques are perfected for it. Pushing ahead with de-extinction by cloning inherently means taking risks.

To recap, animal cells contain DNA in two places: the nucleus and the mitochondria. The latter are the powerhouses of the cell that reside outside of the nucleus, but within the cell's jelly, called *cytoplasm*. Each mitochondrion has its own small genome. So unless either the mitochondrial genome of the animal to be cloned is swapped into the egg cell along with the nuclear genome, or both components come from individuals of the same species, there will always be some—potentially import-ant—genetic differences between the original animal and the new clone. And in cases where a species has evolved to have mitonuclear compatibility, meaning that its mitochondrial genome and its nuclear genome are adapted to each other, that might matter for de-extinction.

Would transferring the nuclear genome of a woolly mam-moth into an Asian elephant egg with its own mitochondria yield incompatibility issues? No one knows yet, but if the answer turns out to be yes, in order to clone the mammoth scientists might need to also retrieve and insert or recreate the full sequence of ancient mammoth mitochondrial DNA. This will need to be investigated for any de-extinction cloning attempt that involves an egg donor from a different species than the one researchers are trying to reanimate. But some of the scientists I've spo-ken with dismiss the importance of mitonuclear compatibility and say that differences between nuclear and mitochondrial genomes won't matter much. Others, however, tell me that they absolutely will matter in de-extinction—it's just that no one knows yet by how much.

The first clone of an endangered species was born in 2001. The nuclei from the cells of an individual endangered bovine—a gaur—were transferred into donated eggs from a cow that the researchers obtained from a local slaughterhouse. Forty-four embryos were made this way and shipped to a company in Iowa called Trans Ova Genetics, which specialized in selectively breeding cattle for the meat and dairy industry. Several of their cows were implanted with the modified embryos, in the hope that at least one would develop into a clone. Eventually a preg-nancy took, and a clone named Noah was born, but he died two days later from dysentery. If he had lived, the plan was to trans-fer him to the San Diego Zoo to become the world's first cloned animal on display.

The second endangered species to be cloned, in 2003, was a banteng, which lived at the San Diego Zoo and managed to survive for seven years. In 2008, clones of a gray wolf were born using cells taken after the wolf had already died. And domesticated cats have been used as surrogate mothers and egg donors for the creation of African wildcat clones, which thrived and could have kittens of their own.

Cloning works reasonably well when you're creating nearly identical copies of species that still exist and have intact egg cells to give. But cloning a species that died out long ago, such as the woolly mammoth, is much more complicated, since the cells will certainly be damaged. Even when beautifully maintained mammoth carcasses that are pulled out of frozen Siberian soils ooze with liquid resembling blood, the cells that scientists recover have always been damaged. As a result, scientists need to instead use the egg cells from a closely related living species, like the Asian elephant, destroying the dream of one day bringing completely genetically identical mammoth clones to life.

In a 2014 UK Channel 4 television documentary, *Woolly Mammoth: The Autopsy,* a group of international researchers is shown pulling an extremely well preserved mammoth—named Buttercup—out of the frozen Siberian tundra. She had been there for 43,000 years before they found her. There are shots in the documentary of the researchers plunging knives into her thawing corpse to remove what look like fresh flank steaks—they are wet and juicy, and at one point, someone pokes Buttercup's elbow and a dark gray and brown liquid starts dripping from her body. In broken English, a researcher says, "Today we go barbecue... mammoth," and they all start chuckling. Later, back in the lab, Roy Weber, a professor in the Department of Bioscience at Aarhus University, spins the liquid in a vial to run tests on it, allowing him to see that the dark liquid was indeed full of red blood cells. But they had all broken open and released their

BUCARDO (PYRENEAN IBEX)

hemoglobin—the molecule that carries oxygen in blood—out into the slurry. Time takes its toll on cellular integrity, and it's tough to find cells that are in great shape.

If an entire species is extinct, then an animal that is similar-sized (for ease of birthing) and closely related (for genetic reasons) must be used as a surrogate mother to give birth to the cloned animal. But it can be tricky for a closely related species to bring an embryo of another species to term. The womb that the extinct animal was intended to grow in went extinct along with the species itself, and the substitute is not guaranteed to be a good alternative. A mammoth fetus developing inside a female elephant is not the same as a mammoth fetus developing inside a female mammoth, for example. If a fetus is growing in the uterus of another species, that developmental environment is going to change the way its genes are expressed. There are all sorts of hormones from the mother that can affect the fetus when life starts to sprout. The mother's immune system might reject the implanted fetus, and incompatibility problems could arise—for example, the fetus may be unable to process nutrition from the mother. Not to mention that an elephant is not a mammoth and therefore does not know how to teach the baby mammoth to act like a mammoth. It would be like a chimp teaching a baby human to be human. An elephant mother knows best how to model the life of an elephant for her offspring, so interactions between the recreated animal and the environment it is placed in—which has likely changed since the time of extinction—could create a slew of new behaviors or affect its gene expression in ways that the extinct species never experienced. Then there's diet: the available foods might have changed since the extinct species kicked the can. This altered diet could interact with the animal's microbiome and also affect gene expression, taking the animal even further away from what it is supposed to be.

Despite the unknowns, some have attempted de-extinction with cloning and come remarkably close to success. Take the case of the bucardo, also known as the Pyrenean ibex. The bucardo was a wild mountain goat that lived in the Spanish Pyrenees until the year 2000. That year, the very last one, a thirteen-year-old named Celia, was crushed to death by a falling branch. But before Celia died, wildlife veterinarian Alberto Fernández-Arias had taken some cells from her ear and frozen them in liquid nitrogen, a method of freezing called *cryopreservation*. This kept Celia's cells whole and intact—exactly what is needed for cloning.

For some years following Celia's death, reproductive physi-ologists led by José Folch transferred the nuclear DNA from her frozen cells into the eggs of a living goat, which had had their own native nuclei removed. The goat eggs filled with bucardo DNA were then implanted into surrogate hybrid ibex and domes-tic goat mothers; it took more than two hundred implantation attempts in fifty-seven goats for seven of them to get pregnant. Out of those seven pregnancies, only one was brought to term, and on July 30, 2003, Celia's clone—the world's first unextinct animal—was born, weighing in at 4.5 pounds. Roughly ten minutes later, though, she died; an extra growth on her lung pre-vented her from flourishing, and the bucardo became extinct a second time around.

When the biological setbacks of cloning are set aside, a glar-ing question remains: Was Celia's clone really the world's first unextinct animal? From the outset, Celia's clone was not a pure species clone but an interspecies clone, born from an egg cell and developed in the womb of another type of goat. Does the hybridity of her bits make one think differently about her than if she'd been created with the building blocks of a single spe-cies? Even more puzzling, does cloning a single animal count as a true de-extinction, or is a species only unextinct when a whole population exists or, better yet, when the population can repro-duce on its own? If a reproducing population is what counts, at what point does it go from being a group of individual animals to the legitimate aimed-for species? Do the animals have to enjoy a certain quality of life that the extinct species had for their de-extinction to be seen as a success? Certainty, to say the least, is not the selling point of this science.

GENETIC ENGINEERING

In 1953, with the aid of Rosalind Franklin's X-ray crystallogra-phy, James Watson and Francis Crick revealed the double helical structure of DNA. It was a watershed discovery. After some following experiments, scientists came to understand how DNA has transferred genetic information in all life forms since the last common ancestor, 3.8 billion years ago. The central dogma of cell biology tells us that biological information is transcribed from DNA into a related molecule called RNA, which is finally translated into protein. When it is in protein form, the proteins carry out the many cellular functions that make us and all other creatures tick.

Some of the earliest experiments in genetic modification involved mutagenesis techniques that blasted organisms with X rays to create mutations. DNA naturally gets scrambled when it is bombarded by radiation, for X rays change the genetic structure deep inside cells. The first of these experiments were conducted in the 1920s, and they picked up steam in the 1960s as more people went to town trying it out. By radiating hundreds of plants and insects, scientists found many new varieties. Those we liked would be cultivated, which was done with rice, barley, oats, bananas, tomatoes, peanuts, pears, lettuce, and so many more. Ruby Red grapefruit is a natural mutant, but with time its flesh lost its depth of color and became a more subdued tone of pink. So scientists radiated the Ruby Red grapefruit to see if they could come up with something more lasting. They created a redder variety this way, one in 1971 that they called Star Ruby, and another in 1985—Rio Red.

As time went on, scientists learned how to move genes around inside an organism more precisely than what's possible through the randomness of sex and radiation. Eventually, it became possible to look inside cells to where the DNA is stored and to read out, or "sequence," the genetic information encoded within those cramped spaces. In the 1970s, scientists figured out how to directly alter the DNA molecule: these first recombinant DNA experiments were carried out largely by Paul Berg,

Stanley Cohen, and Herbert Boyer, who recombined genes from different species of bacteria into the same DNA molecule and, in doing so, revolutionized the manipulation of genes. Nearly half a century has passed since then, and in the interim we've evolved a rather sophisticated culture of genetic designers—scientists, students, and biohackers—who intentionally pass genes back and forth between and within species in whatever ways they wish.

Today, computer servers around the world hold genetic databases that in turn hold the sequences of thousands of organisms in the form of binary code. We've become adept at reading the long digital scripts of As, Cs, Ts, and Gs, the genetic maps of a wide range of biological beings. These databases, which are mirrored around the web, can be used to resynthesize their contents. In other words, we're not only reading but also *writing* DNA molecules by making real genetic strands in the lab that we design using digitally stored sequences. The ability to create DNA from online databases opened up the possibility not just of fabricating genes on demand, but also of combining genes in novel ways that don't exist in nature.

Although it's easy to overstate how radical such technologies are, our ability to read, modify, design, and write genomes is a turning point in the evolutionary history of life on Earth. And it is this perspective—physical biology as binary code and back again—that has slid de-extinction along the likelihood continuum from the far out toward the feasible.

GENE EDITING

Cloning made big strides in the 1990s when Dolly arrived. Then, in the 2000s, when molecular biologists found a way to make artificial enzymes called *zinc-finger nucleases* that can target specific places in the genome and replace or delete the genes that they find there, a type of genetic engineering known as *gene editing* ignited new excitement. As that technology was developing, another enzyme system that could do the job even more quickly, TALENS (transcription activator-like effector nucleases), came on the scene. While both were great tools, neither system was without problems. When scientists sent instructions into the cell about where the enzyme should make an edit along a DNA strand, the edit might be made where it was supposed to, but often it missed the mark. As a result, it could take months or even years before experiments started to work. If only something easier, more precise, and cheaper would come along! Then, conveniently, it did.

CRISPR

The first reference to a natural biological system that would trump other types of gene-editing tools in accessibility, efficacy, and cost was made in 1987 by Japanese researchers who accidentally discovered its underlying machinery in *E. coli*. At the time, they didn't know what they had stumbled upon but described its peculiar features so that others might follow up on the discovery.

Inside the genomes of bacteria, the researchers had found repeating stretches of DNA that looked like palindromes—when they were isolated and sequenced, they read the same left to right, and right to left—and that also had chunks of DNA that did *not* look like palindromes (that is, they were unsymmetrical) slotted between the ones that did. More than a decade later, researchers at Danisco, a Danish yogurt company, observed that some bacterial strains in their experiments were surviving viral attacks, and that these strains shared something curious in common: the same repeating palindromic DNA sequences that the Japanese researchers had described. It took another fifteen years, and the minds of many more researchers, before the powers of this phenomenon as a gene-editing tool would become widely known.

CRISPR stands for "clustered regularly interspersed short palindromic repeats." These recurring identical symmetrical sequences of DNA evolved in bacteria and archaea (a domain of microbes) as a way to ward off infection from invading viruses. When a virus infects a cell, it injects its own DNA into it. The viral DNA then hijacks the host cell's machinery for replicating genetic material and causes it to create lethal amounts of viral particles that break free from the cell, killing it. But when researchers analyzed the DNA that they found between the repeating palindromic sequences in CRISPR-containing microbes, known as *spacer DNA,* they discovered that the code of the spacers resembled viral DNA. It wasn't just a coincidence: the spacer DNA *was* viral DNA.

You can think of the spacers as what evolutionary biologist Eugene Koonin calls "mugshots" of viral DNA that the microbe tucks away in its genome for easy retrieval. When a suspicious visitor arrives, machinery in the cell, encoded by the repeating sequences, uses the mugshots to identify whether the visitor poses any threat. If the DNA on the mugshot matches the DNA of the visitor, then the cell knows it is dealing with a virus that had attacked it sometime in its ancestral past. A matching mugshot signals for an enzyme called Cas9 to arrive on the scene and, with the inspecting machinery, to form a structure that acts like a pair of molecular scissors. With their combined strength, the CRISPR-Cas9 complex chops up the invading viral DNA, stopping the infection dead in its tracks. (Other enzymes can also work with CRISPR, but Cas9 is the best understood.) If it happens to be the first time that the microorganism has been infected by a particular virus, the CRISPR system is also able to store bits of the cut-up viral DNA as spacer DNA in its own genome for protection from future infection.

In 2012, collaborating scientists Emmanuelle Charpentier, from Umeå University in Sweden, and Jennifer Doudna, from the University of California, Berkeley, demonstrated a way to program these molecular scissors to cut up not just the DNA strands of naturally invading viruses, but any DNA they wanted. By inserting a custom strand of RNA into the cell with the CRISPR system—a molecule that can be programmed to be complementary to the DNA sequence one wants to cut—the cell is cleverly tricked into finding the DNA sequence it is told to. When the RNA strand finds the right place, the CRISPR-associated enzyme attaches to that place, changes shape, and then cuts the double-stranded DNA molecule.

In addition to cutting the DNA, CRISPR can insert *new* genetic material, depositing DNA into the genome at the site that has just been cut to effect a desired genetic change. For example, a strand of DNA that codes for blue eye color can be cut out and replaced with DNA that codes for brown eye color. If you place a new bit of DNA that includes the genetic change you want into the general vicinity of the cut site, the cell will try to repair itself by inserting that DNA at the point of the break. This is how CRISPR became the blockbuster gene-tailoring machine it is today. As stem cell scientist Paul Knoepfler says in his book *GMO Sapiens,* CRISPR is like a Swiss Army knife that includes a magnifying glass for finding the right place to cut, scissors for snipping the DNA at the site, and a pencil for writing new code where you want it.

Another remarkable thing about CRISPR is that it works in all systems, not just bacteria and archaea: it can be used to edit genomes in plants, animals, and even humans. Scores of researchers are now developing CRISPR techniques to repair genetic mutations that cause debilitating diseases and fatal illnesses, such as sickle cell disease, muscular dystrophy, and various cancers, as well as to engineer beneficial changes in genes that could have therapeutic effects on cells. Given all its potential, CRISPR has been hailed as one of the biggest biochemical revolutions of the last hundred years, if not *the* biggest. But many scientists tell me that it is still somewhat error prone and doesn't always work as planned. One of its dangers is that you might have a pair of erratic molecular scissors cutting up DNA in places it wasn't meant to, sending cells into an unintended frenzy and possibly causing disease. However, efforts are being made to perfect the technique, and a lot of progress has already been made to restrict possible mistakes.

Although most investments are being made in the technology's development because of its potential for human medicine, and even agriculture, CRISPR is also being used in much more frivolous applications. How about a wee gene-edited "micropig" as a family pet, for example? Its creator, the Beijing Genomics Institute, sells the teeny, tiny genetically modified companion at $1,600 U.S. a pop. And CRISPR's name lends itself to some really bad jokes. I read the following paper solely for its name: "CRISPR Bacon: A Sizzling Technique to Generate Genetically Modified Pigs."

CRISPR is also cheap. Whereas zinc fingers once cost around $5,000, the DNA synthesis required to make the RNA strands that guide the CRISPR machinery inside cells can be as inexpensive as $30. We're talking about an inexpensive, increasingly precise, and possibly universal platform for editing genomes.

In the past, gene edits were made one at a time, but CRISPR allows multiple changes to be made in one swoop by inserting several guide RNAs into the cell at the same time. This is how George Church and his team at Harvard smashed the gene-editing record in 2015, when they used CRISPR to simultaneously edit sixty-two locations in pig embryos, effectively inactivating retroviruses that were known to prevent pig organs from being transplanted safely into humans. Pig parts were already being used in reparative surgeries on human hearts, but those parts have had to be reduced to their underlying structures: their surrounding cells have been removed to prevent the human body's immune system from rejecting pig tissue. They serve mainly as architectural support. However, Church's work indicates that perhaps one day we'll be able to use full, fleshy pig organs in humans when human donor lists fall short of demand.

And when it comes to making the million-plus edits needed to modify the genome of an Asian elephant so that it comes close to matching that of the woolly mammoth, a tool that can make several edits at once is a gift. Indeed, CRISPR's potential for genetic rescue has not gone unnoticed. As Stewart Brand writes, "In parallel with the arrival of 'precision medicine' for humans, where treatment can be specific to the genomes of individual patients (and even individual tumors), we might see the development of 'precision conservation' techniques based on minimalist tweaking of wildlife gene pools." When an endangered species with low genetic diversity is on the brink of extinction, Brand points out, CRISPR could be used to insert key genetic variance into its genome.

PREVENTING DANGEROUS APPLICATIONS WITH CRISPR

As CRISPR ushers in a new era of cheap and easy genome design, precision medicine, and genetic rescue, many fear it is also rolling out the welcome mat for dubious use. The much-feared "designer baby" who will grow into an athletic star with the muscle tone of Michelangelo's *David* and an unbeatable brain is often mentioned as a reason for a CRISPR crackdown. Would you prefer a tall child or a short child? A brunette or a blond? How about a kid with perfect pitch? Or even one that glows green under ultraviolet light?

In April of 2015, in a world first, a group of Chinese researchers announced that they had edited human embryos with CRISPR. They were attempting to edit the gene that causes a potentially fatal blood disorder called beta-thalassemia, but their results showed that making heritable changes to human cells this way, while possible, was still quite imprecise. Although the embryos they used to reach this conclusion were purposefully unviable and therefore would never be able to come to term in a pregnancy, their results gave important warning about editing embryos that could grow into real people one day. The researchers cautioned against using CRISPR any time soon to make heritable changes that could be passed down to the next generation.

The Chinese scientists were doing research on gene editing for disease correction rather than prototyping an enhanced human race, but that hardly quelled the controversy about the use of CRISPR for unintended purposes, like eugenics. The publication of their work created an explosion of controversy across a breadth of fields. To some, a CRISPR-filled future for our species looks bright, ensuring a better life for people with genetic illnesses. But to others that future looks like a less tolerant world, with a disturbing taste for efficiency and a lack of vision for the merits of disability. As journalist Michael Specter wrote in the *New Yorker*, "Not since J. Robert Oppenheimer realized that the atomic bomb he built to protect the world might actually destroy it have the scientists responsible for a discovery been so leery of using it."

When a technology is considered dangerous, moratoriums are sometimes proposed to slow down or stop the technology's development or use. A famous historical case demonstrates how moratoriums can sometimes be voluntarily self-imposed. In 1974, the recombinant DNA revolution was launched when a group of scientists came up with new tools that allowed them to combine genetic material from multiple organisms. These innovating scientists were also the first to come up with and apply regulatory measures to their tools. While they could see recombinant DNA's potential benefits for society, they could see even more clearly its possible unintended consequences for the environment and human health. They decided that the responsible thing to do would be to deter anyone from using it until more research on its risks and benefits had been conducted, and so they established a temporary moratorium on the technique. That sounded reasonable, decent—impressively egoless, even. But good gestures don't guarantee good results.

One morning, the snow outside the bay window in my old living room in Toronto was falling directly into the waves of white hair atop George Church's head on my laptop screen. Impenetrable to the flakes where he sat in his office over five hundred miles away, Church told me about his ironic experience with the voluntary moratorium. "I lived through the recombinant DNA moratorium, which actually accelerated my research," he said. As a security measure, his lab had been tasked with building fancy containment facilities full of state-of-the-art equipment. But there were large enough gaps in the restrictions imposed by the moratorium that one could tinker with the technology without violating the rules that the scientific community had agreed upon. Church and his colleagues managed to continue their work on the basic enabling technologies for genetic engineering while still abiding by the moratorium—except now they could do it with better and safer equipment. So when the moratorium was eventually called off, there was a huge boom in human knowledge about recombinant DNA, which made the possibilities that everyone was originally worried about all the more feasible.

"Not only that," he explained, "but it got so much news attention that investors were thinking, 'Wow, this must be a revolution, I've got to get in on it,' and then suddenly there was this huge set of investors who previously had only invested in electrical, chemical, and mechanical engineering who were suddenly investing in this revolution. With recombinant DNA, if the intention was to strap it down and make sure it goes forward in very incremental secure steps, the result was the opposite in my opinion." Then, realizing how this might sound, he added a clarification: "It ended up fine, but it wasn't what was intended."

This all points to one principal issue: "The problem with a moratorium," Church told me, "is not that it slows down scientists but that it is a false security. It gives you the impression that you are slowing things down when you are not. I'm all for regulation, but just like we need to think about unintended consequences of the science, we need to think about unintended consequences of the regulation as well."

Various moratoriums have been suggested for different types of gene editing with CRISPR in human cells, but they have not been applied in many parts of the world. In fact, in the time it has taken me to write this book, I've watched CRISPR evolve from the focus of hyperbolic and reactionary warnings about its promises and perils, to becoming a rudimentary tool used in even the most basic of genetic laboratories all around the world. It is being applied to humans, animals, plants, microbes, and even extinct DNA for which researchers are increasingly perfecting its accuracy. Today, as I'm writing this, scientists in Church's lab are using CRISPR to edit ancient woolly mammoth DNA into living elephant cells. As the debate continues about whether or not there should be a moratorium on using CRISPR in humans to various ends, these molecular scissors are being sharpened in labs like his for much wider application. As scientists become more sophisticated in using CRISPR to move genes around in human cells and from extinct species into living animal cells, they may become more adept at applying it in experiments that some consider less savory.

SYNTHETIC GENOMES

Another ambitious way to achieve de-extinction would be to synthesize or "write" the entire genome of an extinct species from scratch in a lab and import it into a living relative's egg cell that had its native nucleus removed, which would then get embedded in a surrogate's uterus, where it would develop into a baby animal. But no de-extinction research group has ever officially suggested that this should be done for a candidate species.

To date, entire chromosomes have been synthesized only in bacteria and yeast. This is done by means of machines that use chemicals to hook individual genetic letters onto a column, one by one, to create very long DNA strands that can be stitched together into chromosomes. But for de-extinction to work, the synthetic chromosomes would somehow have to then be wrapped up tightly in a nucleus with the right architectural structure and successfully inserted into a recipient cell without any damage being done to them in the process. Scientists do not yet know how to do this, and some big barriers remain to be overcome. So far researchers can make DNA strands that are many thousands of base pairs long, but the process is not yet refined enough to create the billions that would be needed to make, say, a mammoth genome. And there are long, repetitive parts of the genome that are extremely difficult to sequence, let alone synthesize and rejoin. Plus, no one really knows yet how important these repeating regions are or what role they play in an organism.

Another consideration with synthetic genomes is that DNA synthesis is still expensive. In 2016 Church and others announced the Genome Project–Write, which will synthesize all of the more than 3 billion paired DNA bases that constitute the human genome, as well as the genomes of other species, and which is partly designed to lower the cost of DNA synthesis. If the project is ever completed, it could make it a lot more possible for de-extinction to be done this way. But right now, the general consensus is that it's still too costly and difficult to think about reconstructing an extinct species through whole genome synthesis when more feasible methods exist.

Today's realistic de-extinction research relies on selective breeding (artificial selection; backbreeding), cloning, and gene editing. Selective breeding may be the cheapest and easiest way to go if you've got time to tinker with several generations of breeding pairs and don't care too much about precise gene pools. Cloning will get you extremely close to a genetically identical copy of the original animal but raises the question whether the nuclear genome from the extinct species and mitochondrial genome from the host species are compatible, as well as vital animal welfare issues (though all methods of de-extinction will raise animal welfare issues at some point). Gene editing is versatile and allows for a wide range of extinct genes to be swapped into living animals' genomes, but it also creates numerous issues around interspecies compatibility and authenticity. The arbitrariness of the number of genes edited—which scientists decide on—suggests that gene editing may be as much art as it is science. And as with cloning, the embryos, fetuses, and newborns it creates may fail once they enter or exit the womb until the techniques are perfected. Keeping in mind that there's no magic bullet for achieving de-extinction, all of these experiments must begin the same way: everyone adjusting their idea of what it means to make an extinct species reappear in the first place.

CHAPTER TWO

WHY IS DE-EXTINCTION IMPORTANT?

The Geological Timescale

IMAGINE THAT YOU are bobbing out in space. Before you, the Earth pulses with the energy of all of the births and deaths it has experienced through time. Hallucinatory powers allow you to fast-forward and rewind the events of evolutionary history, as though you were scrubbing through a YouTube video. In the surrounding cosmos, in the flash of an eye, a star is born and dies. Up close on Earth, species lines become syrupy, oozing from one creature into another as natural selection does its work, changing the way the creatures look as they evolve. The dinosaurs that didn't die out at the end of the Cretaceous Period sprout wings and fly, becoming beautiful birds. Everything is in constant change; there are no static shapes to be seen.

The Mexican American philosopher, artist, and writer Manuel DeLanda, while noting that everything that matters to evolution happens across millennia, says of this exercise, "The observer would see species mutating and flowing. He would probably worship flows—unlike us, who, because of our very, very tiny time-scale of observation, tend to worship rocks." We like to focus on things in front of us that seem stable in our lifetimes, from buildings to cities and nation-states. Most of us don't look at the Grand Canyon and see it moving, but it is.

Ironically, we talk about the history of all events on Earth, which occur through dynamic motion, with a standard static image: the geological timescale. This essential tool for geologists, paleontologists, and other earth scientists sometimes appears in grayscale but sometimes, to my preference, comes in more psychedelic colors. Some versions appear in standard chart form, while others morph into mountains, animals, or three-dimensional spirals. But all of these varieties follow a simple method of historical measurement. The geological timescale is a chronological record that connects the layers of rock in the ground to time, chunking events that have happened on Earth together. In other words, geological time is divided into slices according to the age of rocks. It's a bit odd to think about, really—that something as ephemeral as time should equal rocks.

Big slices of rock-time are called *eons*, which are subdivided into *eras*, which in turn are divided into two or three *periods*, which can be divided into *epochs*. The system is hierarchical. Specific moments of rapid species evolution and diversification are told through the fossil record, which can be read out from the great rocky timelines that we drill out of the ground. Births, blossoms, renewals, devastations, extinctions—nothing stands still. Yet this beloved hierarchy of stone, in all of its fixed representation, has come to determine how we think about the flows of life through geologic *time*. The geological timescale allows us to see that extinction is an integral part of evolution, and even to visually map when significant extinction events have taken their toll. And one day, if a certain group of scientists gets its way, we will map the arrival of recreated extinct species on this timescale too.

Extinction—the failure of an entire group of organisms to adapt to changing circumstances and thus their completely dying out—feels intuitive to us today as a concept, but it was not always so. The idea is not inherently obvious and didn't even exist with any proof until the end of the eighteenth century, when the French anatomist Georges Cuvier discovered that molars he had pulled out of the ground could not have come from any organism known to humans at the time. Scientists had seen the remains of old, disappeared animals before, but they had always thought these were just geographically shifted remains of the animals they already knew. In their eyes, for example, mammoth bones (or mastodon molars, as Cuvier had found) were just the skeletal bits of elephants that had migrated north.

Today, however, we know that the possibility of extinction looms over all species, even our own. The background extinction rate—the standard and unrelenting culling of living things over geological time (often measured in extinctions per million species-years)—makes constant disappearances seem somewhat mundane. It's estimated that over the last 3.8 billion years, nearly 4 billion species have evolved on Earth; staggeringly, 99 percent of them are already said to be gone. The vast numbers of disappearances are, however, often balanced by the appearances of new species, which slowly fill the cup back up again—which is why that 99 percent feels hardly possible. But it's not an even teeter-totter effect. If you were to use that same rewind and fast-forward function you scanned the history of life on Earth with earlier, you would notice the bright flashing of five moments of extreme species collapse against the background. These outlying moments are known as *mass extinctions,* and they have, time and again, in varying circumstances, aggressively downsized life's inventory.

Mass Extinctions

THE FIVE MASS extinctions occurred near the end of the Ordovician Period and the start of the Silurian (approximately 443 million years ago), in the late Devonian (about 373 million years ago), at the end of the Permian (about 250 million years ago), in the late Triassic and at the start of the Jurassic (about 208 million years ago), and at the end of the Cretaceous (about 66 million years ago). Depending on how they are measured, the extent to which they surpass the normal background extinction rate fluctuates; yet each greatly exceeds the norm for species recessions in any other timespan of the past 540 million years. Theories about their main causes vary—glaciation in the Ordovician, chemical weathering and carbon storage in the Devonian, global warming and changing ocean chemistry at the end of the Permian, volcanic eruption in the late Triassic, the asteroid that knocked out the dinosaurs by the end of the Cretaceous. (Many of the causes are still heavily debated.) It is said that these five moments alone have swallowed up over 75 percent of the species that the Earth has ever seen, a number that is itself hard to swallow.

Mass extinctions are massively diverse. There is no single cause for them, and no conclusive evidence for why or how they happen. But in all cases, it is largely believed that they arise from a perfect storm of factors—atmospheric composition, climate fluctuation, abnormally high-intensity ecological stressors, and so on—meeting at just the right time to tip ecosystems toward devastation. This doesn't account for fluke incidents like a meteor crashing into the Earth, but these factors "prime the pump" of extinction, and any additional disaster may punctuate a collective die-out with extra oomph.

In 1986, a scientist named Jack Sepkoski defined *mass extinction* as "any substantial increase in the amount of extinction (i.e., lineage termination) suffered by more than one geographically wide-spread higher taxon during a relatively short interval of geologic time, resulting in an at least temporary decline in their... diversity." This definition is now only one among many, but it is a useful way to measure. (Sepkoski also said that to qualify as a mass extinction, an event must go further than the elimination of one group of species.)

According to Sepkoski's definition, eighteen intervals along the geological timescale can be said to count as mass extinctions, but only three of those extinction events—the end-Ordovician, end-Permian, and end-Cretaceous—stand out using the criterion of magnitude alone. The other two that together with these make up the big five—the Devonian and Triassic extinctions—are technically mass depletions in comparison. But they too precipitated extinction so significantly that they are considered similarly terrible times.

Today, mammals radiate and rule, but many scientists say that we are again faced with a blinding flash of obliteration: the sixth mass extinction. What is special and unique about this time is that it is said to be the first mass extinction to be driven by a single species—us. According to many biologists, the sixth mass extinction has in fact been going on, in one phase or another, for about a hundred thousand years, since humans started to disperse across different parts of the globe. There is no unanimous verdict on this, though, and whether it is a *mass* extinction event, like the big five, is debated. Researchers do not agree on the quality of the data available or on which parts of it best measure mass extinction. Nevertheless, it is generally believed that pollution, habitat destruction, overhunting, and the introduction of non-native species into various eco-systems over the last two hundred years—since the Industrial Revolution—have been especially brutal. The extinction rates, expectations, and exact numbers of threats and disappearances get muddled across hundreds of research papers, statistical analyses, and reviews. But what's clear is that we must recognize what is happening and make a plan to address it. If humans are thought to have caused the devastation, do we have an obligation to undo the damage by any means if we can? And if we can, do we dare to try?

Determining whether human beings are responsible for extinction events is no simple affair. If one moment on the Earth's twenty-four-hour clock equals about 200,000 years, we *Homo sapiens* did not show up on Earth until the last moment before midnight. The most recent epoch is just a tiny sliver of Earth's history, spanning roughly the last 12,000 years (or 11,700 years with a maximum counting error of 99 years, if you want to get technical). Known as the Holocene, this epoch began at the end of the Pleistocene—the last glacial period—and is characterized by the development of major human civilizations. It includes our gradual transition to modern urban living, with industrialization, consumerism, and the myriad environmental impacts of contemporary life that are now taking their toll on the planet. Climate change, mass species extinctions, and other misfortunes characterize different time periods in ways people don't always agree on. But a lot of what has happened in the last 12,000 years simply isn't debatable. Humans have transformed ecosystems as powerfully as the geophysical forces that used to do the work on their own. It's even made some people wonder whether we're still in the Holocene or in a new epoch of our own making altogether.

The Anthropocene

AFTER THE HOT northern summer of 1988, known as "the greenhouse summer," the World Meteorological Organization and the United Nations Environment Programme established the Intergovernmental Panel on Climate Change (IPCC). The purpose of this organization was, and still is, to review and communicate the effects of climate change, with a hope of informing policymaking based on solid science that clarifies anthropogenic effects. We all know the story that the IPCC has to tell us by now, and many of us are overwhelmed by it. The human population has boomed, our industrial practices have flourished, and we're making the habitats we so desperately rely on, as well as the marginalized communities that live there, pay.

The global human population was around 300 million in AD 1000, 500 million in 1500, and 790 million by 1750—a steady increase. It began to skyrocket at the beginning of the Industrial Revolution, in the late eighteenth century. Now that we have surpassed 7 billion, we've changed the physical sedimentation of the planet, making it impossible for some species to live in habitats where we've eroded the land through agricultural and urban development and have dammed riverbeds with construction sites, to just scratch the surface of what we've been doing. The global temperature is expected to rise around 0.2 degrees Celsius (about a third of a degree Fahrenheit) per decade as a result of human-caused carbon emissions, and many habitats will become unlivable for species that have adapted to inhabit them in cooler temperatures. Although various species might be able to cope with the changes, it's not likely that all will adapt quickly enough to survive. Meanwhile, land developments are obliterating former escape routes, trapping animals in areas where they might not evolve in time to make it.

As the IPCC reports, the oceans have changed too. Increased ice melt has caused sea levels to rise, and as a result of increased carbon release, the Earth's surface waters now have a significantly more acidic pH than in preindustrial times. Increased acidification could prevent several important biological processes from being carried out for a wide variety of marine creatures. What was once a mundane part of sea life—growing shells and skeletons—could become a strenuous, if not impossible, affair. In acidifying waters, huge numbers of mollusks' larvae are expected to fail or develop abnormally along with— and I feel a stinging just thinking about it—the acidification of the mollusks' blood. That type of sanguine suffering will take a toll on how well mollusks can respire and excrete. Scientists have said that the ocean floor will start to dissolve in response as chemical reactions attempt to neutralize the acidifying waters. Although this is not an exhaustive list of possible effects, scientists are exhausting *themselves* trying to come to a consensus about when this degree of devastation really began and what that means for how we should understand it.

In 2000, Paul Crutzen, a Dutch Nobel Prize–winning atmospheric chemist who had made major discoveries about the depletion of the ozone layer, and Eugene Stoermer, a biologist with expertise in freshwater species, proposed a term to describe the outsized influence of humans on the planet that defines our current time: *Anthropocene*. They declared that the planet had entered an altogether new geological epoch, distinctly different from the Holocene. This hypothetical epoch, demarcated by human expansion on and domination of the planet, marks the moment when our species started affecting the Earth's systems to the same degree as did geological factors like plate tectonics and the rise of mountain ranges. Its time of origin is hotly contested, with suggestions ranging from the beginning of the Holocene approximately 12,000 years ago, to the Industrial

Revolution, to just after World War II, when the fallout from the first nuclear weapons created a worldwide radionuclide signature in the Earth's sediment. The jury is still out on when it started and on whether or not it will become an official term in scientific stories about the history of life on Earth.

The Anthropocene may have become an environmental and intellectual buzzword, but some critics point out that the concept misplaces the blame for how we've changed the world. The problem is that it treats humans like one big category—as though it were *all* of us who have created the damaging impacts of the Anthropocene. Of course, that's not the case. It's the wealthiest of us who have done that. And as with climate change, it's the least wealthy among us who first feel its effects.

EXTINCTION IN THE ANTHROPOCENE

Today's sixth mass extinction is similar to the other five in that it represents a giant loss of biodiversity in an incredibly short amount of geological time. According to a 2014 publication in the academic journal *Science,* of the 5 million to 9 million animal species that we know about, which is a conservative guess, anywhere from 11,000 to 58,000 are vanishing per year. Those estimates do not take into account local extinctions, called *extirpations,* in which the species still exists elsewhere. A study published in 2015 claims to show "without any significant doubt that we are now entering the sixth great mass extinction event." The researchers looked at vertebrates, the most extensively studied group of animals, which includes mammals, birds, reptiles, amphibians, and fish, and provided conservative estimates of the number of species that had gone extinct since 1900. Despite their cautiousness, they could still show that 468 more extinctions had occurred than would be expected under normal geological circumstances—a number believed to be around 9. For the invertebrates—land-crawling critters without spines—the data is much less complete. But research suggests that invertebrates are in even more trouble than their bony-backed counterparts. It is estimated that 26 percent of mammals will be wiped out if the current extinction rate continues, and some have said that the extinction rate will increase from 1,000 times as high as the natural background rate—the current situation—to 10,000 times as high in the future.

The impact of the Anthropocene on species extinctions is unclear. On the one hand, some scientists say humans have ushered in the sixth mass extinction by expanding our societies across the globe in the truest Anthropocene fashion. On the other hand, the pressure that the Anthropocene puts on ecosystems also forces new species into existence. Chris D. Thomas, a professor of conservation biology at the University of York in the UK, argues that hybridization and the blurring of species lines, as accelerated by the Anthropocene, need not be maladaptive. Throughout human history we've translocated species on purpose, spread invasive species by accident, and brought formerly separated species into contact with each other by remodeling their habitats. Hybridization flourishes where humans shape the land, affecting the dispersal of animal populations. The Arctic Spring now arrives roughly one week earlier than it used to, and the winter freeze sets in one week later than it once did, but animals compensate for the temperature change by moving into new ranges. Along the way, they might meet other species on the run from melting sea ice and mate with them, creating a generation of new hybrid species. According to Thomas, "Speciation by hybridization is likely to be a signature of the Anthropocene... Populations and species have begun to evolve, diverge, hybridize and even speciate in new man-made surroundings." Climates change, habitats morph, and new creatures come into view.

Stewart Brand recognizes the Anthropocene's creative potential for speciation: "Any creature or plant facing a shifting environment has three choices: move, adapt or die. Evolution is far more rapid and pervasive than most people realise." He points out that evolutionary change does not always mean evolutionary disaster—and climate change doesn't mean that all species will die in its wake. Some might move into new areas beyond their historical distribution, expanding their native range. In this sense, the Anthropocene might devastate some species, but it might also accelerate the evolution of others.

If hybridization in the wild can allow creatures to survive anthropogenic change, what might that mean for hybrid species that are intentionally created? When genes from extinct species are inserted into the genomes of their living relatives, will de-extinction become an asset for the Anthropocene? As other human-discovered technologies like CRISPR and cloning allow the threat of extinction to be alleviated from inside a petri dish, what types of environmental change will the creatures they produce be able to cope with when they're reintroduced in the wild?

The Anthropocene is creative, an idea Brand pushes to its edge. He has argued that mass species extinction is not the problem—the real issue is the way we use its narratives to stir panic in the minds of the public. Our hearts ache for species as they disappear one by one, but, Brand says, the decline of many nonextinct wild animal populations is a much greater threat to conservation than the obliteration of single species. Ecosystems are affected not just by how many species exist in them but also by how many individual animals are there to play out their ecological roles. This is what's known as *bioabundance,* without which a substantial number of important ecological processes—like grazing, planting seeds, and enriching the soil— might happen too infrequently to maintain an ecosystem's productivity. In this sense, the more individual creatures there are, the more ecosystem richness there will be. "Viewing every conservation issue through the lens of extinction threat," Brand says, "is simplistic and usually irrelevant. Worse, it introduces an emotional charge that makes the problem seem cosmic and overwhelming rather than local and solvable."

The philosopher Timothy Morton speaks of *hyperobjects*—entities spread so vastly across space and time that we can't see their edges. Climate change is a pertinent example. Mass extinction also aptly fits the bill. When you can't see where a hyperobject stops or starts, you might not know where or how to intervene. Hyperobjects make us feel small and helpless. In light of extinction, Brand states, "The core of tragedy is that it cannot be fixed, and that is a formula for hopelessness and inaction. Lazy romanticism about impending doom becomes the default view."

As a champion of de-extinction, Brand understandably wants us to feel that it can help us do something proactive about species loss instead of give into environmental pessimism. And by steering us away from the hyperobject at hand and focusing us squarely on pragmatic goals—such as increasing the number of animals out there—he presents the quest for bioabundance as convincingly worthwhile. But do we really want to obliterate all distinction between the *kinds* of species loss we are concerned about and the degree of that loss? When a species goes extinct, a particular way of life is lost forever. Creating new transgenic animals in the name of conservation may be fine in cases that have been thoroughly thought out. But it does not account for the fact that when a certain species disappears, its unique flavor of existence, which had intrinsic meaning and value to other life forms, fades away as well. What would we miss if bioabundance were all that mattered? Perhaps if bioabundance was our foremost concern, we'd risk undercutting the moral value of living species and all that their existence has brought into the world so far. Call me old-fashioned, but why shouldn't that matter more?

REVERSING EXTINCTION IN THE ANTHROPOCENE: "THE MORAL HAZARD"

"I am terribly sad that you are writing a book on de-extinction," Stuart Pimm, a prominent conservation biologist with ruffled gray hair and a direct way of speaking, tells me barely one question into our interview. That's because he smells a strong whiff of human-centered egotism in de-extinction, which, he says, "sounds very sexy, but also sounds very white-men-wearing-lab-coats-who-are-going-to-save-the-planet." His disapproval is rooted in a worry that we, as a public, have become overly attuned to stories of environmental doom and gloom. He fears that the extent to which we mire ourselves in dire narratives about the end of biodiversity leaves us without any room for stories of hope—or *real* hope, as he puts it, since he doesn't think de-extinction fits into that category. "What I would love for you to write a book about instead is how enormously successful conservation has become at saving species," he tells me. Yet Pimm was one of the first scientists to show that species are disappearing one thousand times faster today than they normally would otherwise. How is that hopeful? He knows the facts and knows they're grave. What gets him out of bed in the morning, he says, is that he gets to work with brilliant people all over the world to come up with practical solutions for saving species. However, he will never count de-extinction among them.

His voice tenses as he scoldingly says, "I think it is a real shame you are not writing about real solutions. Is de-extinction a solution? It's a tiny, microscopic marginal solution with a lot of problems. There's a lot of fantastic people out there doing incredible things to save biodiversity, save species, but this is not anywhere near the top of the list." And he's right: there are a lot of conservation success stories to tell that we don't seem to celebrate in society nearly enough. For example, Pimm's NGO, Saving Species, has helped take the golden lion tamarin—a tiny fire-orange monkey from coastal Brazil with a crotchety face—off the endangered species list. With the help of local conservation groups, the NGO purchased 270 acres of cattle pasture that separated two dislocated areas where the golden lion tamarins live. The populations were choked on either side of the divide, unable to meet and procreate. But when the cattle pasture was turned into a monkey thoroughfare, the separated tamarins—which were nearing perilous isolation on both sides—could breed and flourish across the land.

Before that, Pimm was involved in the rescue of the Florida panther, a species that had become so badly inbred that the males were suffering all manner of reproductive nightmares—they had low sperm count and low sperm motility, and in many cases their testicles wouldn't drop. At one point in the 1990s, no more than thirty panthers lurked in the Florida Everglades, so conservationists introduced eight female Texan panthers into South Florida to see if they would create hybrid kittens with the floundering males. It was a controversial step, and even Pimm was skeptical at first. But the new blood made for a breathtaking turnaround. The males, despite their limitations, eventually managed to do what was expected of them, and hybrid kittens were born. Those kittens grew up three times more likely to make it to adulthood than purebred kittens and as adults expanded the Florida panther's range. Those are the kinds of stories that Pimm wants us to take hope from in these times of mass extinction, not some "crock aspiration" to restore ecosystems with unextinct animals that are not even authentic replicas of the species they seek to replace.

Another problem, in Pimm's eyes, is the moral hazard of assuming that we can pull de-extinction off without a hitch. "The issue is the moral hazard of saying, frankly, we can drive species to extinction because we can always bring them back." In his work, Pimm encounters a lot of people who, he says, would like to drive species to extinction, at least at the local level, for their own financial gain—people who want to develop valuable animal habitats and remove the populations within them, by promising to return the species (the ones that their own commercial projects would displace) at a more opportune time. For example, he has testified before U.S. congressional committees in cases where the multi-billion-dollar logging industry wants to chop down vast amounts of old-growth cedar, fir, hemlock, and spruce forests in the Pacific Northwest but are held back from logging legally because of a pretty, white-spotted brown bird that lives there.

In 1990, the spotted owl was declared a threatened species, and conservationists restricted how close loggers could get to its nesting grounds. But Pimm has since witnessed a number of proposals that suggest protecting the bird in puzzling ways— for example, by cutting down significant amounts of its habitat and ushering the owls into remaining parts of the forest. If the birds don't fare well with the changes, it's been suggested, they could be put in captivity and returned to the wild when the forests have regrown. But they could also just be left where they are in the first place. Pimm believes, based on evidence like this, that before any ecologically beneficial new animals are created, de-extinction will introduce a moral hazard by suggesting that it doesn't really matter if we drive a species to extinction because we can always bring it back at a better time.

Ryan Phelan doesn't think so. "The moral hazard thing just doesn't resonate for me," she says. She tells me that the same idea came up over thirty years ago when the first frozen zoos were established to bank cells and tissue samples from endangered species. At the time, people thought that frozen zoos—large repositories of species-specific DNA archived in sub-zero temperatures—would cause people to assume that nature will be protected against all odds. But the fact that vials of endangered species' DNA are maintained in frozen zoos has never been an excuse not to protect the living version in the wild.

Another concern is that money for conservation programs could be diverted to biotechnological solutions like de-extinction. An attempt to remove support for species protection was observed when people started cloning endangered animals, like Noah, the gaur, in the early 2000s. In fact, shortly after Noah's arrival was announced, critics of the Endangered Species Act suggested that species will no longer go extinct, argued for an overhaul, and, in some cases, called for complete removal of protective legislation for endangered species. Then again, de-extinction and conservation programs aren't likely to reach into the same purse for funds. In many cases, de-extinction efforts are looking for donations from benefactors who would like to see specific species return. The idea that a wealthy donor dying to see a mammoth-like elephant come to life would otherwise care about helping the endangered birds of Hawaii with his or her own private money is a bit of a stretch. In other words, a donor's money for de-extinction may end up in the pot for a particular species or in no pot at all. Yet, for every species that is brought back, there are going to be thousands that there will not be the money, time, or interest to reanimate. So if de-extinction ever becomes largely driven by deep pockets and species favoritism, what will that new biotechnologically mediated wilderness look like? And, importantly, how much will it cost?

Conservationist Joseph Bennett of Carleton University is the lead author on a 2017 paper in *Nature* that caused a flurry of debate about the economics and possible related damages of de-extinction. In "Spending Limited Resources on De-extinction Could Lead to Net Biodiversity Loss," Bennett and his colleagues set out to determine how the relative costs of establishing and maintaining populations of resurrected species would pan out in terms of their impact on conservation programs for endangered extant species, without factoring in the direct costs of the experiments needed to make the unextinct animals in the first place. They concluded that if government-funded conservation programs were left to pick up the costs in order to ensure that unextinct species have a shot at making it in the wild, even after private donors may have paid for the creation of the animals, there would likely be considerable costs to currently endangered species. Based on realistic baseline conservation budgets in New Zealand and the Australian state of New South Wales, they argue that conservation programs would receive less support, and thus, as a direct result of de-extinction, net biodiversity would decrease. Considering New Zealand's conservation budget, they calculated that if 11 unextinct species were to receive conservation funds for maintaining their resurrected populations there, that nearly three times as many living endangered species could be maintained on the same amount. "If taking care of these resurrected species becomes the purview of governments, then something's got to give unless somehow the budgets go up a lot and the governments can afford both the resurrection of species and the extant endangered species they should be working on as well," Bennett told me.

But Revive & Restore has stated many times that they are looking for private funds, not government funds, to back their projects. And this raises the convincing point already mentioned that those private funders might only care about seeing their money resurrect an extinct species, or nothing at all. In their paper, Bennett and colleagues also consider a scenario in which private funders pay for the whole shot of setting unextinct populations up in the wild and managing them there. Their numbers show that private funding for five unextinct species in New South Wales could instead be used to conserve over eight times as many endangered species (42) there. They warn that even when programs are privately supported, there are critical missed opportunities since the funds could have always gone to endangered species conservation instead—costs that outweigh the benefits of de-extinction.

"If an agency wants to see a mammoth and wants to look one in the eye, then that agency is not necessarily going to spend the money on conservation," Bennett said, in agreement with what Revive & Restore have proposed. "But my problem is when de-extinction gets couched as conservation. The people who say, 'Oh, we'll bring back the mammoth, but certain agencies aren't interested in spending money on other endangered things,' are the same people you will also see couching mammoth resurrection in terms of being conservation. If it is conservation, then they can do a better job with those resources, so couching it in terms of conservation is disingenuous. But if some agency is like, 'Look, we want to make a mammoth and look one in the eye,' I'll say okay, fine. I mean, I'd love to look one in the eye. It would be really cool. But I think you can make one point or the other point. You can't make both."

Ben Novak, Revive & Restore's lead scientist on their passenger pigeon de-extinction project, criticized the paper for conducting its analyses based on unrealistic candidate species that no one is currently planning on "de-extincting." He also pointed out that Bennett et al. assessed the costs of existing conservation programs in a small and very particular part of the world (New Zealand and New South Wales) that is not representative of the settings most de-extinction projects will occur in.

When discussing costs through another lens, the official guiding principles of the International Union for the Conservation of Nature (IUCN) in *Creating Proxies of Extinct Species for Conservation Benefit* point out that ensuring successful translocation of unextinct animals to release sites and properly caring for them once they're introduced into their selected habitat will require considerable costs in the form of "not just money but scarce human resources." As a field, conservation isn't exactly teeming with professionals who are qualified to do this type of work, so allocating those people to work on de-extinction projects rather than more traditional conservation translocations or management projects could cost endangered species in material ways that reach beyond just the bottom line of conservation budgets.

To this end, some critics wish that de-extinction advocates would simply call a spade a spade and get on with their future-forward show. Rather than argue for the conservation merits of de-extinction or our moral responsibility to pursue it, they urge us to celebrate how neat de-extinction is in a whizbang kind of way. "I think it is a very cool project technologically, but most of the environmental reasons people use to justify why we should do it are silly or wrong," says Tom Gilbert, a professor of paleogenomics at the University of Copenhagen and director of a large laboratory for ancient DNA research at the Natural History Museum of Denmark. Forest elephants in Africa are in grave danger from heavily armed poachers with tremendous firepower who gladly kill conservationists and wardens standing in their way. Why are we sitting around talking about woolly mammoths if we don't even know whether we can keep the forest elephants alive?

The western black rhino was declared extinct in 2011 as a result of poaching. Two of the remaining five species of rhino—Javan and Sumatran—are critically endangered in Asia, and the northern white rhino, a subspecies of white rhino, is extinct in the wild, with only three living individuals remaining at the Ol Pejeta Conservancy in Kenya. In South Africa, the number of rhinos killed has risen exponentially—while only 13 were poached in 2007, a record-breaking 1,215 rhinos were maimed in 2014. What if paying more soldiers to protect rhinos in the wild turned out to be cheaper and more effective than remaking other beloved extinct species and returning them to the wild? "If we really care about animals going extinct," Gilbert says, "let's stop trying to bring the damn things back and let's try to spend the money on things that are actually going extinct."

Putting a price tag on de-extinction is tricky. Even well-worn conservation programs often keep their books far from view. Pimm's group spends a great deal of time trying to find out how much money is being spent on conservation overall around the world, but his two words for anyone trying to track down the numbers are "Good luck." Conservation programs are staggeringly expensive and can be remarkably complex. They take a lot of time and effort, and they're not a one-time investment. At a certain point, you're going to have creatures roaming around in the wild that need to be managed, and someone is going to have to pay for that work.

In 2015, when I asked Phelan how much Revive & Restore spends on de-extinction, I was happily surprised by how openly she responded. "I'm getting tired of it looking like we have raised all this money and have zillionaires backing this," she said. "You know, over the three years that we have been working in this area, we've raised and spent about a million bucks." Most of that money had been spent on public education, private workshops, the TEDxDeExtinction program, outreach, and some of the science. Revive & Restore pays for the sequencing costs and lab materials that some of their researchers require, but that doesn't add up to very much. For de-extinction and genetic rescue projects to really move forward at a faster rate, since 2016 Phelan has been engaging commercial labs (Crystal Biosciences and Dovetail Genomics) and biotech companies (Intrexon). Her goal is to raise $1 million to $10 million each year, through either donations or in-kind services. After that, lifelong monitoring of each species, for several generations of lives, needs to be factored in.

It seems that they're off to a good start with their fundraising, with over twenty-five donors listed on their website as founding funders who have given $10,000 or more to their cause. I thought it quite fitting to see the name of one of the most famous living fantasy writers on that list—George R.R. Martin, author of the novels that form the basis of HBO's hugely successful television series *Game of Thrones*. De-extinction attracts creative minds with imaginative ways of seeing the world, but it remains to be seen how its stories will come to life off the page.

CHAPTER THREE

WHAT SPECIES ARE GOOD CONTENDERS, AND WHY?

(Re)Born to Be Wild

"MAYBE WE CAN edit long-dead genomes back to life. Maybe extinct species could walk the Earth again. Maybe they could once again thrive in the wild," Brand once wrote in an online debate about the merits of de-extinction. His opponent was Paul Ehrlich, an esteemed conservation biologist who considers de-extinction a "fascinating but dumb idea." Notice that Brand didn't write, "Maybe they can teach us about genetics," though surely they could, or "maybe we can visit them in a zoo." He wrote, "Maybe they can once again thrive in the wild."

Traditionally, a reintroduction occurs not when a species is extinct but when a population of that species no longer exists in a particular area where it used to—animals from the surviving population are moved to that area to fill in the gap. This is more broadly known, in biological parlance, as a *conservation translocation* and can include more radical actions, such as putting animals in completely new areas, outside of their indigenous range.

With de-extinction, on the other hand, a reintroduction requires replacing animals that have vanished from the wild with new ones that have been created to mimic them. The journey starts wherever the animals are made, birthed, or reared—a captive-breeding facility in most cases, sometimes a lab. Their destination might be the extinct animals' original habitat—an ideal scenario. Some argue that unextinct animals must *only* be introduced into the extinct species' indigenous range. But if it were deemed unacceptable to move unextinct animals into an entirely new habitat in the name of de-extinction, would it be okay to go ahead and reintroduce them into the extinct species' ecosystem even if it has changed drastically since the species died out?

"Resurrecting a population and then re-inserting it into habitats where it could supply the ecosystem services of its predecessor is a monumentally bigger project than recreating a couple of pseudomammoths to wander around in a zoo," Ehrlich writes in his debate with Brand, pointing out that even though it might be possible to reengineer extinct animals, there is no guarantee that they'll be able to live well in the wild. There's a saying in conservation biology—credited to the plant ecologist Frank Egler—that goes a little something like this: *Ecosystems are not just more complex than we currently think, they're more complex than we can think.* In the teeming web of life that extends beyond our backyards, we cannot claim to know where all of the nooks and crannies lie. New interactions could emerge in the wilderness between predators, prey, microbes, and a changing environment—interactions that can't be foretold. The variables quickly become complicated. Animals have been successfully reintroduced into the wild many times before, but getting the animals there isn't what's most daunting: keeping them alive there is.

Dolly Jørgensen, at Luleå University of Technology in Sweden, is an environmental historian who studies animal reintroductions. The first time we "met," I could just barely make out her short dark brown hair and glasses through the poor Internet connection in my friend's loft above a Montreal mattress store. The second time we spoke I couldn't see her but heard her Scandinavian accent through the headphones I was wearing while we chatted between radio booths on either side of the Atlantic.

Jørgensen believes that if we are going to do it right, we have a lot to learn from the animal reintroductions that have already occurred and that to overlook these would be a grave mistake. Her research has focused on reintroduction projects in Norway and Sweden, which she studies to understand how past human–animal relationships influence the lives of reintroduced species in the present. "Reintroductions are complicated because you need to properly evaluate why the species died out," she says. For example, you need to first solve the problem that caused the species to decline in the first place. For a reintroduction to be successful, there needs to be no sign left of the pressures that originally killed the animals. That might be easy to understand in theory, but it's hard to execute in practice. Crucially, you need to figure out if suitable habitat for the species still exists. Has the environment changed so much since the species went extinct that a recreated species might not prosper there now?

Take the Yangtze River dolphin. In 2006, after a six-week visual and acoustic survey of the entire range where the species was known to have lived, it was assumed to be extinct. Researchers on the hunt for any sighting of the last thirteen known individuals found nothing. The Yangtze River dolphin was the first vertebrate in fifty years known to have gone globally extinct, driven over the edge by local fisheries and polluted waters. Given the polluted state today of the river it is named after, proposing de-extinction of this dolphin would be ludicrous. If a population were to be recreated now and put back into the river, it would likely be only a matter of time before the dolphins floated back up on shore.

We are always in a waltz of nature and culture, organism and environment, action and reaction. From the right insects to optimal soil types, an ecosystem's function depends on a full choreography of factors, none of which act in isolation. Human beings sometimes lead the steps—when an animal goes extinct because we hunted it, for example. Before we bring that animal back, we've got to figure out why it was hunted to extinction. Did we profit from killing it? If so, then the recreated animal must no longer have that value to us. We also need to know if people were ever afraid of the species. Did it have sharp teeth and a taste for flesh? We have hunted bears, lynx, and wolves to local extinction before because we feared they would kill our livestock and, by extension, our own financial gain. These sorts of interactions need to be known before any de-extinction project can get off the ground. But it takes time to figure these things out. When beavers were reintroduced in Scotland, it took ten years of discussions before a five-year pilot project that included reintroducing a mere four mating pairs could be started. The biggest part of preparing for a reintroduction is trying to convince people not to fear the animal, Jørgensen says, "or else they're just going to kill it again."

She's seen this happen, or come close to it, many times before. To her mind, the worst thing that could happen would be for an unextinct animal to kill someone in the wild. Extreme as that may sound, it is not impossible. Take the case of the muskox, which was reintroduced to Norway in 1932. Muskox had enjoyed the Norwegian way of life for several thousands of years before they went extinct there. When they were eventually reintroduced to the mountains in central Norway, people accustomed to hiking in that area had to quickly adapt to their presence. Conflicts arose when hikers turned a corner and ran into a muskox that would chase them up the stream or scare them so much that they'd fall over and hurt themselves. By the 1950s, as the herds grew and lone males would wander from the pack, these incidents had become more common.

One day in the 1960s, a man in his seventies was out on the trails when he was startled by one of the imposing beasts, which knocked him down. When the man died from his injuries, the local community went into a state of furious outrage, screaming that they would shoot all the muskox in central Norway. "They probably would have, too," Jørgensen says, "if the government didn't get in the way. I can definitely picture some societies where they would have taken out all of their shotguns that day." She fears that something similar, or worse, could happen if a re-created woolly mammoth ever rammed into a truck on a trail and killed a passenger inside. "Then all we will see is outrage."

Although that scenario seems outlandish, it could occur. Andrew Torrance, a professor at the University of Kansas School of Law, tells me that with the way the law is now in the United States, an owner of an unextinct mammoth cannot be prevented from releasing it on his or her own land. So hypothetically, thanks to their strength and size, recreated mammoths may be capable of getting near and hurting humans.

Other de-extinction candidates are not nearly as threatening, however. Take the case of the gastric-brooding frog. At worst, one might lick us and we wouldn't like it. The gastric-brooding frog is believed to have gone extinct in 1983 by way of a perilous fungus: a chytrid that goes by *Bd*, which humans spread around the world and which is still causing global amphibian extinctions. While it was alive, that frog became famous for its ability to incubate its fertilized eggs in its stomach and then vomit up the resulting tadpoles and spew them into the world. It was the only species known to be able to take one of its organs—its stomach—and change it into another organ, a uterus, on demand. Medical researchers became interested in understanding how the frog controlled and regulated this process such that it did not digest its eggs while they were incubating in its belly. But the frog was discovered only in the late 1970s, and just as the medical community's excitement about it was rising, it disappeared. That's partly why some researchers badly want it back.

Michael Archer is a paleontologist at the University of New South Wales and leader of the Lazarus Project, a de-extinction effort that is trying to clone the extinct gastric-brooding frog and that takes its name from the Bible's Lazarus of Bethany, whom Jesus brought back to life. After six years of attempts, in 2013 Archer's team made a breakthrough when they took frozen genetic material from the extinct frog and transplanted it into the nucleus of a related living frog—the great barred frog—to create embryos that carried gastric-brooding frog DNA. To everyone's excitement, the great barred frog cells successfully replicated the extinct frog's DNA. And then they stopped. None of the embryos survived more than a few days. But Archer's team had reached the first crucial step in bringing a clone of the species to life. One of the challenges is the natural reproductive cycle of the great barred frog, whose egg cells are required in order to create the cloned embryos. Because the frog produces viable eggs only one week a year, researchers must grab them at just the right moment. But harvesting them is difficult, and there is only that tiny window of time each year when they can try.

Is it really the right time to clone the gastric-brooding frog when the chytrid fungus that first killed the species continues to annihilate amphibians around the world? Even if the fungus could be controlled, Jørgensen wonders if there might be issues

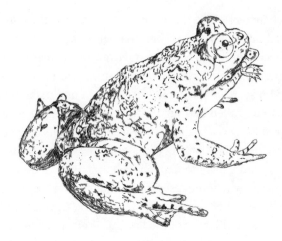

GASTRIC-BROODING FROG

of water quality and pesticide use that could affect a recreated frog's porous skin. If a population of recreated frogs were ever to be released and survive, would they need some form of legal protection from certain chemical industrial practices? Going out on a speculative limb, she imagines a world in which companies are denied their ability to work in an area because it is being turned into a gastric-brooding frog sanctuary. If so, she says, "Somebody loses in this conversation. We have to recognize that it raises conflict." Who is the loser going to be? In many reintroductions, it is not so much that the species are particularly contentious but that the whole scenario is.

Certain Troubles with Certain Species

ALL OF THIS leads to another question: How should we try to manage human–animal conflicts? Dolly Jørgensen believes that it is imperative that we discuss the relationships we want to have with recreated species before any appear. In order to get past solely technical conversations about reintroduction, we need to identify what species, including our own, need. Environmental humanities scholars, like Jørgensen, study the values that are embedded in our stories about species and how those values affect their lives. Some of the more interesting stories to look at, she says, are the ones that we learn as children, which her research shows still matter when we're old. "What we've seen in wolf reintroduction work is that those stories—the fact that it is the Big Bad Wolf that eats Little Red Riding Hood—matter to how people perceive reintroduced wolves."

Take, for example, the thylacine, which went extinct in 1936. Often described as a doglike carnivore with tigerlike stripes, the thylacine was neither dog nor tiger but a carnivorous marsupial that, like a kangaroo, carried its developing young in its pouch. Michael Archer, the same scientist behind the gastric-brooding frog de-extinction project, has shared some of his zoological affection with this species. At the TEDxDeExtinction event in 2013, Archer spoke about the earliest specimen of thylacine that has ever been found by humans—a 25-million-year-old fossil recovered from the ancient rainforests of Australia. He told us how it was discovered amid the remains of marsupial lions, carnivorous kangaroos, "a giant, weird duck" that ate flesh, and peculiar crocodiles that could scale trees. He described, with the zeal of a seasoned storyteller, how the tree-climbing crocodiles would drop down on thylacines like anvils, squashing the carnivores into the ground. The victims varied greatly, from "great big ones to middle-sized ones to one that was about the size of a Chihuahua." He added, "Paris Hilton would have been able to carry one of these things around in a little handbag, until a drop croc landed on her." Over time, as climate change altered the forest, the thylacines began to die out and subsequently lost

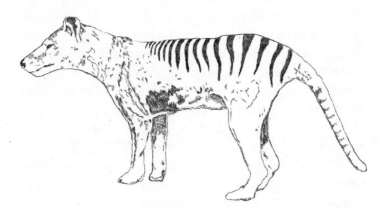

THYLACINE (TASMANIAN TIGER)

some of their vast genetic diversity. All types of thylacines disappeared, except the mid-sized one, which wouldn't vanish until sometime later.

The first scientific description of the species was made in the early nineteenth century, when New South Wales lieutenant governor William Paterson sent a wordy account of its appearance to Sydney for publication in a local paper. Shortly thereafter, Deputy Surveyor General George Harris provided an official description of the thylacine, which was read before the Linnean Society in 1807 and published a year later:

> The length of this animal from the tip of the nose to the end of the tail is 5 feet 10 inches, of which the tail is about 2 feet . . . Head very large, bearing a near resemblance to the wolf or hyaena. Eyes large and full, black, with a nictant membrane, which gives the animal a savage and malicious appearance . . . The whole animal is covered with short smooth hair of dusky

yellowish brown... On the hind part of the back and rump are 16 jet-black transverse stripes, broadest on the back, and gradually tapering downwards... Only two specimens (both male) have yet been taken. It inhabits amongst caverns and rocks in the deep and almost impenetrable glens in the neighbourhood of the highest mountainous parts of Van Diemen's Land, where it probably preys on the brush Kangaroo, and various small animals that abound in those places.

The high-endurance marsupial would outrun its prey until it surrendered from exhaustion and fell right into the thylacine's jaws. It ran across the ancient landscapes of Australia and Papua New Guinea up until about four thousand years ago, when dingoes were introduced to Australia by Asian seafarers. The two doglike species may have competed for the same habitat and resources because of their similar body size and meat-eating habits, and it wasn't long after the dingoes arrived that the thylacines disappeared from mainland Australia. After that, they were found only in Tasmania—an island state of some 26,400 square miles off the southern coast of the Australian continent, called Van Diemen's Land at the time—which explains the animal's second name: the Tasmanian tiger.

Act 2 of the thylacines' decline began in the early nineteenth century after Europeans arrived in Tasmania, bringing weapons and sheep with them. After taking a look at the jaws of the Tasmanian tiger and hearing its yips near and far, the newcomers were quick to assume that their lambs would all be eaten by the bloodthirsty carnivores, though the idea that Tasmanian tigers were aggressive predators of sheep is now believed to be wrong. Recent studies propose that they could not have been that dangerous to sheep at all because they didn't have the right type of jaw to devastate their populations in large numbers, and the sheep may have been more endangered by local dogs. However, landowners across Van Diemen's Land complained about losing sheep by the day, and in 1830, the Van Diemen's Land Company announced the first bounty for Tasmanian tiger heads: 5 shillings for each male and 7 for each female. The bounty lasted several years, as the incentive to kill them increased. Each dead thylacine became valued at 10 shillings if it was in the company of another nineteen or more carcasses. The government announced its own bounty too, paying out more than 2,000 of them between 1888 and 1909.

As early as 1863, the Tasmanian tiger's extinction was being predicted by the naturalist John Gould, in a volume of *The Mammals of Australia*. Nevertheless, the bounty lived on, and it was not long before the yips of the Tasmanian tiger were no longer heard across the land. Benjamin—the very last one—died on September 7, 1936, at Hobart's private Beaumaris Zoo on a cold evening, when his keepers locked him out of his enclosure all night, not realizing that he was certain to die of exposure. When they found him the next morning, his body was swiftly dumped in the trash, and it took fifty more years for the IUCN to declare the species officially extinct. Although Benjamin's DNA was lost forever when the caretakers cleaned out his quarters, a six-month-old female—preserved by pickling her in alcohol in 1866—eventually found her way into Michael Archer's hands.

In his TEDxDeExtinction talk, Archer explains that back in the 1990s he asked some of his geneticist friends if they might be able to extract the DNA from the pup and use it at some point to resurrect the Tasmanian tiger. But they just laughed. This was pre-Dolly the Sheep, and cloning, never mind cloning the dead, was still science fiction in most people's minds. Then, six years later, Dolly arrived, and so much more became possible for people to imagine.

In 1999, when Archer was made director of the Australian Museum, he assembled a research team and embarked on a project to clone the thylacine. Ancient DNA was still emerging as a field and had been surrounded by hype ever since the first DNA extracts had been taken from the quagga skins fifteen years earlier. "Study of ancient DNA was the allure of time travel," according to Svante Pääbo, the scientist who would later publish the Neanderthal genome. Sensing that some scientists were guiding their research according to fantasy rather than legitimate biological questions, Pääbo called out its fatal trap: the hype trap—the same trap that Amy Fletcher, an associate professor in political science at the University of Canterbury in New Zealand, accuses Archer of falling into. She writes, "When it launched the thylacine project, the Australian Museum walked out on the unstable precipice of 'paleogenomics as science' versus 'paleogenomics as spectacle.'" She criticizes Archer for stating that the museum would emphasize the science behind sequencing thylacine DNA and the creation of a genomic library when in reality it spent more energy propping up public excitement around the controversial prospect of resurrecting the extinct.

The science was spectacular in the fullest sense of the word, and unsurprisingly, the media loved the story. An image of the blanched pickled pup curled in on itself like a chubby sickled moon in a see-through glass jar made its way around the world in documentaries and news articles. It quickly became the icon of the thylacine's resurrection. But the alcohol-soaked blob was not the only specimen—or even the best one—used in the museum's studies. In fact, there were three: the pickled pup in the jar, a dehydrated unsexed specimen, and another dried male specimen. The disembodied specimens were far less captivating than the pup and harder to parade around in the global show-and-tell about the thylacine's return. In the 2010 article "Genuine Fakes," Fletcher explains that a museum employee, Allen Greer (who wrote about the project in 2009 from the inside), had pointed out that the pickled pup yielded no impressive amount of DNA that was useful for the thylacine's resurrection. Nevertheless, it remained the poster child for the project, deceiving audiences around the world about the scientific process involved. "At the outset," Fletcher writes, "Archer referred to the pickled pup specimen as 'the miracle bottle in which this time capsule is just waiting to pop back into life' and suggested that 'at the rate at which this technology is increasing, I wouldn't say there's any reason why we shouldn't expect to be able to go into a pet shop and buy a pet thylacine and bring it home.' " (I've asked Archer multiple times for an interview to get his side of the story but never heard back from him.)

When they weren't allegedly working the press, Archer's team toiled away in the museum lab and eventually managed to get some good DNA samples from their specimens. They extracted enough samples from the liver to try to sequence the entire chromosomes, but when they amplified the DNA, they found what many paleogeneticists often do—contamination from the DNA of other species. The contaminants could have come from any variety of microscopic creatures, or even from the researchers who had dipped their hands into the jar over the years.

By 2005, the thylacine cloning project was so endangered by the lack of good available DNA and reported lack of research facilities that it too went extinct. However, the mission never left Archer's mind. When I heard him speak in 2013, he was advocating again for the thylacine's return and trying to convince us that the Tasmanian devil, a distantly related species, would be a good surrogate mother for the task. But Tasmanian devils are currently suffering from an infectious affliction that's killing them in droves, even though Brendan Epstein and colleagues' 2016 study reported that they have been developing some resistance to the cancer called devil facial tumor disease (DFTD)—so using Tasmanian devils in de-extinction would require an awful

lot of management. At the time that I'm writing this, no significant funding has been put aside to revive the thylacine cloning
project . . . though that could always change at any moment, with
the right donor, of course.

There is a video on YouTube from 1933 of Benjamin running
around in his cage at the Beaumaris Zoo, giving us a vivid feeling
and flavor for what this species must have been like. In black and
white footage, he saunters, scratches his hindquarters, and even
curls his lips as he gnaws on a bone. That video collapses time
and creates a portal through which the mistakes we made in the
past creep forward into the present to haunt us. Dolly Jørgensen
tells me, "It is a very emotional story about the reintroduction
of carnivores, which always have a bad reputation, but also
other creatures, like lynx, for example. Lynx are very subdued
and don't kill much livestock, but what we see with their reintroductions is that they are inevitably contentious." Lynx are
flesh-eaters, but they tend to prey on smaller animals such as
porcupines, rabbits, and birds.

Jørgensen has examined reintroduction proposals for lynx in
Scotland and found that people simply don't want lynx around
where they or their livestock are. It doesn't matter if you tell
them that a lynx doesn't kill very often or that it mostly hides in
the woods. Lynx seem threatening because of the stories we've
always been told about animals with big, flesh-tearing teeth. The
same would apply to the Tasmanian tiger: even if all the studies
now show that a thylacine won't eat livestock, cultural narratives
about carnivores that describe them as scary, rogue wild beasts
are still circulating. So a wise use of research funds might be
to look at the successes and failures of carnivore reintroduction
projects *before* anyone tries to recreate extinct meat-eating animals. "Maybe these problems have solutions with the right type
of management," Jørgensen says, "but they need to be thought
about now."

A species reintroduction might sound good on paper, but how does it sound when *you* are the person who has to deal with the consequences of a reintroduced species living near your home? Jørgensen tells me that people often have a "not in my backyard," or NIMBY, attitude toward these things. The NIMBY problem is even well documented in reintroductions that have been considered successful, like beavers, which were reintroduced in Sweden just decades after they went locally extinct. Jørgensen says that all of the beavers in Sweden were gone by the 1870s, while just a few pockets of populations remained scattered around the rest of the continent. In 1922 some beavers were brought back to Sweden from Norway for the first time. Forty breeding pairs were released over the next twenty years. Today there are over 100,000 beavers in Sweden—so many that there's even an annual beaver-hunting season from October to May. People are generally still happy to have beavers around, but conflicts arise when a beaver builds a dam that backs up a town's water onto a farmer's field or onto a road bridge—people don't like inconvenience or extra work. When that has happened, communities apply for permits to remove the dam, which inevitably means killing the beavers. "It's not a conflict that's happening at the whole-species level," says Jørgensen, "but when it is on your property, you don't like beavers very much. So from looking at what's happening with local populations of beavers in Sweden, you start to see the kind of relationship we are really in with animals, no matter what our conservation mandates might say." Human acceptance, rather than technological hurdles, may be the ultimate challenge in de-extinction.

The ways we talk about animals can have multiple effects. For some people, de-extinction and reintroduction is justified by missing species that we wish were still around and that we feel guilty about if humans caused their disappearance. Jørgensen's findings cause her to believe that the act of mourning species, on personal, emotional, and cultural levels, is the oldest motive for reintroductions we have. "In my work on beavers brought back to Sweden," she says, "I see that people thought that it was wrong that they were missing—there was this idea that the Swedish nature was not as good without beavers. They were nationalistic about it. They told stories about their grandmothers using beavers for this and that, and thought, 'Oh, isn't it sad that we don't have it anymore.' So I think it is perfectly natural that de-extinction scientists talk about missing species, and say that we are lesser for them being gone. This is very human, this is not the animal speaking for itself. It is extinct, it can't think. It is *us* saying we want it back. That makes it our decision to do it, and we need to think long and hard about that." Our knowledge of a species' natural history, causes of its extinction, habitat, and behavior, as well as—crucially—our own desire to "have it back," are as vital to making a de-extinction project real as is the funding and technological know-how.

I worry that if we don't learn the sort of cultural lessons that researchers like Jørgensen have to offer, we might end up making lab curiosities or animals for our own entertainment, ones people pay to see in captivity. If any aspect of the cultural context is overlooked, an unextinct species could wind up just as doomed by our interventions as the extinct species was without them. When I ask Jørgensen to tell me how she has been able to inform the debate so far, she says, "I would love to, but historians aren't being asked... It seems that the discussions that are happening are being primarily dominated by natural scientists, and sometimes they talk about culture and people, but that's not what they actually study. They study ecosystems or genetics. And I think it's unfortunate because there are many people who have been doing a lot of thinking about what the human–animal relationship is really like."

Although that may have been truer when de-extinction first hit the headlines, in my own research I've discovered that many nonscientists—from humanities scholars to artists and designers—are now actively weighing in on the debate. The questions about their relevance then become: What spaces are they invited into? Should historians make suggestions for the regulation of unextinct animals before any of those decisions are finalized, or are they better suited to offering pithy commentaries after the fact? As Jørgensen says, "What is de-extinction going to mean to local human communities? Is it going to be accepted? And most of all, how are people going to understand this? That's where we need to contribute." The moment at which their form of "outsider" knowledge is integrated into the dialogue matters, not just whether or not they speak up at some point. There may indeed not have been enough researchers like Jørgensen involved early enough, but that, now, seems to be changing.

Deciding on Candidates

PHILIP SEDDON, A conservation biologist at the University of Otago who was involved in drafting the IUCN's global guidelines for species reintroductions, writes, "Love the idea or hate it, the rapid development of technology that raises the prospect of de-extinction, whatever that turns out to be, will... 'reframe our possibilities.' " And he, like Jørgensen, wants to enrich the Wild West of de-extinction with what we already know about how to manage reintroductions well. In order to pave a path for dealing with the quandaries that de-extinction creates, Seddon—along with two other scientists, Axel Moehrenschlager and John Ewen—came up with some questions to be answered before any de-extinction project is developed. Their criteria are based on the IUCN's guidelines for endangered species management and are designed to bring trusted conservation suggestions for species translocations into the de-extinction fold.

In the paper, they explain that if de-extinction is going to work—meaning that a recreated species can survive on its own in the wild—it isn't necessary that the species be released into the extinct species' native habitat. Instead it might be released into a habitat where the extinct species never lived but where the recreated species can play a valuable ecological role and enrich the wilderness. This could prevent the (re)extinction of an unextinct species in cases where the original habitat is no longer suitable. This type of introduction is usually accompanied by assisted colonization, in which a species is monitored in the habitat until it can exist there without human help.

Based on the preexisting IUCN reintroduction guidelines, the researchers came up with ten questions to ask about any candidate for de-extinction. They suggest that each candidate must satisfactorily pass the following question round before more refined scrutiny comes into play.

1. "Can the past cause(s) of decline and extinction be identified and addressed?"
   Are we sure that we know what factors led to the species' extinction, and are we certain they are no longer a threat? Given that 99 percent of the species that have ever existed on Earth are said to now be extinct, many or most extinction events will have been poorly documented. In these cases, it can be tough to infer what the direct and indirect biological, physical, social, political, and economic factors led to the species' demise. When assessing the safety of bringing a poorly documented species back to life, we can get lost in the fog of best guesses. But if the extinction record is thorough, it can be used as a barometer for determining if the culprit

that caused their extinction will still affect the species should it ever reappear.

2. "Can potential current and future cause(s) of decline and extinction be identified and addressed?"
If we recreate a new version of the species, could it face current or new threats that might propel it toward re-extinction? Although it is crucial to understand what killed a species in the first place, it is also important to think about other factors that could wreak havoc on an unextinct animal. Given what we know about how habitats may have changed, are there any future threats that a species might face at release sites and in the surrounding habitat? If threats exist but seem manageable enough for the plan to proceed, that could still require engineering the entire ecosystem's dynamics to an even greater degree to ensure that the threats don't become too large.

3. "Are the biotic and abiotic needs of the candidate species sufficiently well understood to determine critical dependencies and to provide a basis for release area selection?"
Do we know enough about the species' natural history to be able to determine what the animals absolutely require and to find them a habitat where they can flourish based on those needs? This is where historically inclined researchers get to have the most fun. Only after having pored over oral histories, expert reports, field notes, and experiments can we say we know a sufficient amount about the living and nonliving factors that an extinct species once depended on. Information about where its populations were distributed, how its social systems operated, what it ate, how it reproduced, and how it cared for its young is vital. When confronting prehistoric species like the woolly mammoth, we have to go hunting

for clues in its DNA. But in cases where humans recorded detailed observations of these creatures, we have much more to go on.

4. "Is there a sufficient area of suitable and appropriately managed habitat available now and in the future?"

Is there enough suitable habitat for the species? And can we keep it that way? Selecting an appropriate release site (the actual geographic coordinates) and release area (the wider region where the animals are expected to disperse and settle in) is vital for reintroduction. We know from conservation science that translocations often fail when unsuitable release sites are chosen. The longer the animal has been gone, the less likely it is that we'll have complete information about the animals' needs and its ecological role. We need to know if the identified release area is too small for a fully restored population, is unavailable, or is unprotected.

Here, a de-extinction decision making body must weave a tight braid out of what we know about the animal's natural history and the on-the-ground reality of the available habitat. To further complicate matters, regulators must also assess the potential of that habitat to endure future climate change fluctuations and possible human-caused land use perturbations. Such an assessment will help regulators understand whether the habitat might be restored in a way that does not threaten the return of a viable population. For example, if the release site or release area contain trees that industrial interests hope to clearcut one day—and they have a real chance at doing so—it would not be ideal to reintroduce a species there.

5. "Is the proposed translocation compatible with existing policy and legislation?"
Will putting the species back into the wild be legal and advis-
able? In some cases, because of technical demands, the lab
work needed to revive a species might take place in countries
other than the one where the animals are to be released. In
that case, international agreements and conventions would
apply, adding an extra layer of bureaucratic negotiation to
already bumpy terrain. Depending on how the species is
revived, it may be subject to different patent laws in different
parts of the world or may be eligible for endangered species
protection according to a contextual kaleidoscope in which
all geographic distinctions are tossed together in a complex
legislative and policy-based potpourri. Chapter 6 deals with
these sorts of legal questions.

6. "Are the socioeconomic circumstances, community attitudes,
values, motivations, expectations, and anticipated benefits
and costs of the translocation likely to be acceptable for
human communities in and around the release area?"
Are locals willing to deal with the economic, environmental, social, political, and cultural costs of bringing recreated
animals to live in their region? Where a resurrected species
carries cultural, economic, or human health value, people
who live near their release site will have their own interests
in mind. For example, some might find it beneficial to their
social status if they could serve braised heath hen at their
next dinner party. Others might want to start a business that
brings nature-tourists to the area where unextinct animals
live, charging money for the rewilded view in order to bring
wealth to their community or themselves, as we've seen
with True Nature Foundation's intentions to backbreed the
aurochs. Conversely, having the species back could prevent

those same people from operating businesses in certain areas or making money in particular ways. Stakeholders' opinions and community attitudes toward introduced species need to be solidly understood.

7. "Is there an acceptable risk of the translocated species having a negative impact on species, communities, or the ecosystem of the recipient area?"
   Once the animals are back in the habitat, are they going to mess it up for the other creatures that live there? Determining what's acceptable in that respect is a daunting task. Could bringing a species back put other species in peril? Could an introduced species hybridize with creatures that are already living in the release area, making the extant ones genetically morphed beyond recognition, and would that be a bad thing? We must strive to make the risk-to-benefit analysis of species and their habitats as clear as can be.

8. "Is there an acceptable risk of pathogen-related negative impacts to the resurrected species and the recipient system?"
   Are there hidden pathogens in the extinct species' genome that might be recreated along with the species itself? Or existing pathogens that the species might be susceptible to in the release area above a level of acceptable risk? We have seen that when an endangered species is translocated to a new area in the hopes of avoiding a known pathogen, that pathogen can simply reappear in that area. Take the case of the endangered Tasmanian devil, which is susceptible to a pathogen that causes a fatal face tumor. The pathogen spreads when the animals bite each other, so theoretically, if the tumor is not yet in a population of devils, they should be safeguarded from it. Based on this reasoning, healthy Tasmanian devils without tumors have been translocated

to new habitats and then raised in isolation. Nevertheless, cases have been documented where the original pathogen spread into the new habitat. This species still suffers, just on new soil.

9. "Is there an acceptable risk of direct harmful impacts on humans and livelihoods, and indirect impacts on human ecosystem services?"
Will resurrected species hurt our communities? Will we want to hurt theirs? Do introduced species compete with us, either directly or indirectly, for clean water or food? Do they erode our soil? Do they pollinate plants we don't want them to? Do we like the way they fertilize the land? Will they want to eat our livestock? The list of questions in this category goes on and on. If *we* aren't happy with how an introduced species behaves, they won't be happy either. And if failed assessments of candidate species mean that species can go extinct *twice,* de-extinction could enable a new paradigm of extinction.

10. "Will it be possible to remove or destroy translocated individuals and/or their offspring from the release site or any wider area in the event of unacceptable ecological and socioeconomic impacts?"
If we need to, can we get rid of what we've created? The idea is heinous: Let's go to all the trouble of reviving an extinct species in order to kill it again on purpose. But in the case of emerging danger, we need to know that we can pull the plug. For our own security, and that of ecosystems at large, we have to know that we can bring this experiment to a halt if need be. This would probably say much more about our own poor judgment and planning than about the species itself, but as we all know, humans make mistakes.

Charismatic Species

SEDDON FIRST GOT involved with drafting that set of questions when de-extinction (under the banner of Revive & Restore) hit the headlines in 2013. At the time, the state of the discussion about candidate species took Seddon by surprise. As far as he could tell, people weren't thinking all that hard yet about which species might warrant candidate status, so he read as much as he could to try to understand de-extinction's underlying goals. It struck him that many of the early candidates on Revive & Restore's website were charismatic species with widespread popular appeal. They were the usual suspects—beautiful birds and iconic mammals. "Even in the kind of conservation work that we already do today, there's a taxonomic bias," Seddon says. "We tend to work on birds and mammals. We don't work on other things nearly enough." He sensed that the same species favoritism he knew from traditional conservation was being repeated now in de-extinction, and that concerned him a great deal.

After Seddon published the suggested criteria for selecting de-extinction candidates, the IUCN got in touch with him and explained that the organization was under some pressure from people who were starting to ask what the IUCN's official stance on de-extinction was. But none of the IUCN staff members were even thinking about it yet. So they asked Seddon to head a task force to create the official IUCN global guidelines for de-extinction, the IUCN SSC *Guiding Principles on Creating Proxies of Extinct Species for Conservation Benefit,* which is what he was busy doing when we spoke.

One of the first times Revive & Restore raised the idea of a de-extinction wish list was at a private two-day meeting it organized at National Geographic headquarters in October 2012. That meeting brought together thirty-six scientists from around the world who had been working with technologies that might assist in the mission. Tom Gilbert was one of the scientists in attendance—the professor of paleogenomics at the University of Copenhagen and Natural History Museum of Denmark who is critical of de-extinction's espoused conservation motives. The first time I visited him at work, I was heavily distracted, not only by his pleasant British accent and the Obama paraphernalia that fills his desk (including bobbleheads, Obama-themed play money, and a collection of campaign mugs), but more so by the air rifle over half my size, leaning against the wall to my left. When I nervously asked him about it, he explained that he sometimes takes it with him to the Copenhagen Zoo. I didn't know what to make of his answer, but he swiftly elaborated, saying that he has never aimed it at any of the caged creatures. Instead, he sometimes points it up at the sky and blasts away until it starts raining rocks—rock pigeons, that is. Rock pigeons are the common city pigeons you know well. "They're feral flying rats," he tells me. And their DNA comes in handy for his experiments from time to time.

Gilbert's focus is on innovating methods for extracting and assembling ancient genomes. As a result of his specialty, he was invited to make a presentation on the technical limitations of de-extinction at Revive & Restore's first private meeting, which he happily did, though he feels now that what he had to say might have made him unpopular with the resurrection hopefuls in the room. "I find de-extinction fascinating," he says. "Really fascinating. I just think there's a lot of bullshit wrapped up in it too." He is skeptical about people's motivations for it more than any of its technicalities. To his mind, de-extinction is an understandable pursuit for someone trying to overcome the elaborate scientific and technical challenges of working with ancient DNA but quickly gets laughable when the conversation turns to candidates that are completely ridiculous to think about recreating. He remembers that at the meeting "everyone had their favorites to bring back. Some were charismatic, while others were clearly stupid, like the Steller's sea cow."

The Steller's sea cow was a herbivorous marine mammal that could grow up to 30 feet long and that is outlived today by a group of much slighter sea cows two to three times smaller. "Growing a Steller's sea cow inside of its closest living relatives would be like incubating a poodle in a Chihuahua," Gilbert told me. The puzzle pieces just don't fit. So at the meeting, Gilbert came up with what he thought was a more practical candidate: "Why," he asked, "don't you start with the extinct Christmas Island rat?"

The Christmas Island rat, also known as *Rattus macleari* or Maclear's rat, comes from the same genus as *Rattus rattus,* one of the most heavily studied mammals of all time. Gilbert thinks the Christmas Island rat is an intelligent choice because of how much we already know about the genes in the rat system, making it more likely that scientists would get what they might want working from it. A gamut of tools already exists for editing rat genomes, and rats breed very, very quickly. "Besides," he says, "not that many people get upset when rats die for science. It's a lot easier than killing a lot of mammoths until you get the process to work." Despite the pragmatism of his suggestion, the idea didn't fly at the gathering. But funnily enough, "rats with wings" did, as Revive & Restore's flagship de-extinction project, the Great Passenger Pigeon Comeback—explored in chapter 5—can attest.

The charismatic quality of an animal not only influences which ones people may want to revive but also plays a role in which ones we let slip away. No one understands that better than *National Geographic* photographer Joel Sartore, an exuberant and affable man with chocolate brown hair and an ear-to-ear smile who keeps good cheer despite the dire scenarios he deals with every day. For the past several years, he's been racing to photograph every single species held in captivity around the world before those species completely collapse in the wild. When I met Sartore in 2013, he'd been at that project, the Photo Ark, for seven years and had already managed to capture 3,200 of the 12,000 or so species that live in captivity on the planet.

"That bird is always in the back of my mind," Sartore says, referring to a picture of Martha, the last passenger pigeon—the very reason that he got into conservation photography in the first place. He still remembers the day his mom brought home a *Time Life* picture book with Martha on the cover, when he was just eight or nine years old. When he took the book in his hands, his eyes fixed on Martha and he was unable to move for several minutes. When we spoke, he had just returned from a trip to the Cincinnati Zoo, where the little stone pagoda Martha died in has been preserved. Inside the pagoda, he found himself again just standing, staring.

Before he started the Photo Ark, Sartore had been photographing a lot of snakes and turtles—basically, anything that would just sit still—but the photos weren't resonating much with audiences. So he started photographing anthropomorphic species, ones that have big beautiful eyes and look back at you with a spark of intelligence. That's when people started to care about his photographs. "Now I photograph species that look like they are thoughtful, conscious beings. I'm photographing things that look almost like people now."

Some people condemn the extra attention we give animals that remind us of ourselves and that are therefore easier to empathize with. But Sartore doesn't think that focus is necessarily bad for conservation, even though it could be that many species—the ones that lack anthropomorphic features—will be overlooked. He tells me we should take advantage of the people power that particular species have and just accept that other species never will have that power. For twenty-two years he had traveled around the world taking pictures of forests being cut down, animals that had been orphaned, and the horrors of the bush-meat market, but those photos just did not move people. "This is my last gasp to get people to look these things in the eye. I'm standing, if you will, in front of a ravine where the bridge is washed out and I'm waving my arms at this train coming at us, saying, 'The bridge is out! You'd better slow down!' " And if it takes privileging one species over another in order to put on the brakes, then, he says, so be it. "I don't have to tell people what to think; I just want them *to* think. Is that too much to ask?"

Charismatic species are tricky. On one hand, they make playing favorites far too easy and allow many species to be ignored while only a select few get our compassion. They invite us to overlook the many types of value a species may have and focus instead on how they make us feel. The value of species has been broadly conceptualized along three intersecting axes: intrinsic, biocentric, and anthropocentric. If you recognize the intrinsic value of a species, you see its worth in and of itself. If you recognize its biocentric value, you see its ability to fulfill an ecological role. And if you recognize its anthropocentric value, you see its value to you or other humans—economically, aesthetically, emotionally, or otherwise.

In some cases the fact that charismatic animals make us feel anything at all may be the only thing that saves them. Do you think we should be embarrassed about that? Or instead, work to leverage it? Sartore's work has shown that it can be easier to move mountains than to get people to care about biodiversity if the species doesn't touch hearts and minds. Charisma, we all know, is a gift with great influence, but it doesn't affect all people the same way.

CHAPTER FOUR

WHY RECREATE THE
WOOLLY MAMMOTH?

It Would Be Cool!

PEOPLE ARE OFTEN willing to pay for something that they think is cool: a flashy convertible, a table at a Michelin-starred restaurant, an overnight stay at an exclusive hotel. But what does it mean when the cool thing is a sentient being? What unexpected forms of commodification might that create? Who profits from making its coolness available? And what happens if it goes out of style?

When considering de-extinction's potential applications, conservation consultant Kent Redford and colleagues write, "The work will attract funding, inform science, help develop techniques useful in other fields, and provide an example of synthetic organisms that have public appeal." But that already raises an ethical issue: Should we be promoting the public appeal of synthetic organisms when we could be working harder to increase the public appeal of unmodified creatures that are still around and that face threats in the wild? Are these two things mutually exclusive? Or would it be a mistake to think that they are?

In 2015, *Jurassic World,* the fourth film in the Jurassic Park series, earned over $200 million at the U.S. box office on opening weekend. In 1993, the original film's opening weekend roped in almost a quarter of that amount. It would seem that the story about undoing extinction only grows cooler with time. "It would be awe inspiring to look at a woolly mammoth walking around... It would be cool. It would be like that first time I turned the corner and saw Yosemite Valley spread out before me," says Hank Greely, a professor of law and (by courtesy) genetics at Stanford University who has been actively engaged in the de-extinction debate since Revive & Restore first gave it a boost. Certainly, it would be remarkable, memorable—dazzling, even. It would be something we were glad to have seen. It would be cool to hear the noises that a woolly mammoth makes with its trunk. It would even be cool to see how it sleeps. But would it be cool to see it slink its trunk through the bars of a zoo enclosure? Would it be cool to watch it amongst a herd of elephants, the only hairy one of its kind? Would it be cool to see it led around by its caretaker for the third time in one day while families of onlookers gawk and wave in its direction?

When Greely says that it would be cool to see a mammoth, he is not imagining that grim scene. He's thinking about how seeing a mammoth could reconnect people to the natural world they are a part of through the emotion that its reappearance would likely create. "And that is a real advantage," he says. "Most of what we do in our lives, we do because we hope for something cool at least, awe inspiring at best." I can easily imagine that a cage full of passenger pigeons or thylacines would also generate a huge amount of enthusiasm. Some people would do anything to see them. Even the paleobiologist Tori Herridge, of the Natural History Museum in London, who deeply condemns the idea of mammoth resurrection on ethical grounds, says, "For all my protests, I'd pay to see one if it was there."

On the contrary, Hendrik Poinar, who grew up thinking about de-extinction and now directs the Ancient DNA Centre at McMaster University, says that his nightmare scenario for de-extinction would be if we ever bring back target species so that people can pay money to visit them in a zoo. He is adamantly against de-extinction if human entertainment is the outcome, a not improbable prospect in his eyes. "I don't think that the people who are at the grassroots of this have that in mind," he tells me, "but it is also silly to think that it won't get out of their hands very quickly. I mean, all you have to do is control the patent. Scientists have seen that in the past. They think they have control of the situation, but it's really the people with the deep pockets that do." He knows from personal experience that making money from de-extinction theme parks isn't just material for a Hollywood fantasy but also an enterprising goal for some people. A wealthy businessman once tried to lure Poinar away from his academic position over a lunch that included a seven-thousand-dollar bottle of wine and an offer to join his private effort to bring the mammoth back to life. The businessman imagined that revived mammoths could live in a park north of Toronto, where he'd bring a slice of the Pleistocene back to life for tourists to enjoy.

Nothing went forward with that plan after Poinar turned him down, but others have raised the idea that animal re-creation could open up profitable businesses. Researchers have described the recent rise of "last-chance" tourism, where customers pay to visit near-extinct animals and landscapes before it's too late. Patrick Whittle, Emma Stewart, and Davis Fisher hypothesize that based on the interest in so-called last-chance tourism and the public awareness of disappearing nature that goes with it, de-extinction could provide a "first chance" for customers to see particular kinds of nature appear, building up a market around "re-creation tourism."

When viewed pessimistically, such ideas sound as if de-extinction may become yet another technological platform that allows capitalism to get its claws into nature. But the idea of making money from de-extinction may not be in competition with de-extinction's ecological merits, just as putting a few revived animals on display may not be incompatible with reintroducing many more of them into the wild. A species' revival will be tremendously costly, especially in cases where bespoke technologies will need to be developed or advanced. If even a handful of unextinct animals are exhibited in a zoo, for example, research funders might be able to recoup some of their costs or even generate revenue to make the next species revival possible. Such an exhibition might also have great scientific merit. As Ryan Phelan tells me, "I think that there is no question that zoos will play a role in assisting any kind of captive breeding that would have to happen before the release of any species." She sees zoos as an intermediary that can help smooth the animals' transition from the lab into the wild. If the specimens on display are framed as a cultural curiosity rather than as a scientific oddity, then exhibiting them might cause people to have more of a stake in the animals' well-being. Would you pay to see the first members of an unextinct species return? And how would seeing them in captivity make you feel?

Mammoths Would Keep It Cool

PUTTING MAMMOTHS' COOL factor aside for a moment, what might the real, measurable benefits of recreating woolly mammoths be? One argument de-extinction advocates make for having unextinct mammoths is about elephant conservation. The idea is that by editing the genomes of elephants with select woolly mammoth genes, we will make cold-tolerant elephants that will allow today's currently endangered elephants to live in wider ranges. Due to the mammoth-like features that gene editing will endow them with, they will be able to colonize vast northern ecosystems, which they currently cannot. This conservationist viewpoint requires that recreated woolly mammoths or cold-tolerant elephants, however you choose to see them, should return to the cold habitats where the original woolly mammoths lived, such as Canada's Yukon and the vast Siberian steppe. But could they become acclimatized to those landscapes today given the fact that the planet is warming? Conservationists are already worried that the polar bears in the North may soon have no sturdy ice left to stand on. So why should we think that new cold-tolerant animals would have a better fate? I was astonished to discover that the fact that the North is warming is exactly why some people are saying that we need the woollies back there now. The campaign for their resurrection relies on an ecological argument about climate change that posits that the woollies could be tools that can help *keep* the North cold.

PLEISTOCENE PARK

Russian scientist Sergey Zimov often has a tough, stoic expression on his "outdoorsman's" face. He also has a big, bushy beard, which is most visible when his long honey-gray hair is pulled back in a low ponytail. For most of the year, Sergey and his son Nikita work out of a remote post in northeastern Siberia, where their scientific research center, the Northeast Science Station, has stood since 1977. The station lies a few miles outside Chersky, a town with a population of roughly three thousand, a quarter as many as before the Soviet Union collapsed.

When you first arrive at the airport in Chersky, "it's a real shocker," says Max Holmes, a senior scientist at the Woods Hole Research Center in Massachusetts who studies carbon fixation in the permafrost found in that part of the world and who has made several trips to the Zimovs' station over the years. He describes the closest airport as "a junkyard for airplanes—and then there is not much except for empty buildings along the roads until you reach the station. It's an edge-of-the-world sort of place, and Zimov is a one of a kind for sure."

The Northeast Science Station is a forty-five-minute boat ride from a large wilderness area called Pleistocene Park that Sergey Zimov founded himself. The park is a giant rewilding experiment that is trying to return the ecosystem to what it was like over ten thousand years ago. It's all part of an audacious plan to save the world from a climate disaster, but it started humbly in Zimov's backyard. The area of Pleistocene Park makes up just a tiny part of the vast mammoth steppe, which dominated the Arctic in the late Pleistocene and spanned Europe, northern Asia, and northern North America. The Zimovs are working to recreate the mammoth steppe ecosystem inside their park by introducing species that can fill the ecological roles active during the last formal epoch. Sixty percent of the park is floodplain, which provides shrubs and grasses for the introduced animals to eat. The rest is beautiful boreal forest of sparsely dispersed larch, surrounded by more than a million lakes. In the open-water seasons, when the river is thawed, the only way to get around in the area is to float up the tributaries.

During the Pleistocene, the mammoth steppe was as rich in wildlife as the African savanna is today. But as the climate changed and the number of marauding hunters increased, the animals began to disappear. By counting the bones in the ground of the park, Zimov estimates that there was once one mammoth, five bison, six horses, and ten reindeer for each square kilometer (about 250 acres), with some muskox, elk, woolly rhinos, saiga antelope, snow sheep, and moose peppered among them. The predators that preyed on those creatures—wolves, cave lions, and wolverines—increased the animal density of the area even more, to ten tons per square kilometer. Grasses dominated the land in a coadaptive response to the animals' constant grazing, while most of the trees and shrubs were trampled under their hooves.

After lifetimes spent feeding and defecating, the animals died and fell into the ground. The mass of accumulated organic material from their feces and bones built up over thousands of years, contributing to the carbon-rich soils found there today. Yet the remains of the animals and their effluence make up only a tiny fraction of the total carbon in the soils. The more substantial deposits come from the dead plants they fed on. As time passes, the frozen collections in the ground expand toward the surface and an icy glaze grows over the roots and bones, which are piled high beneath the ground. These developments, over millennia, have created enormous stores of carbon in the soil. If the permafrost ever rapidly thaws and these carbon stores are released into the atmosphere, scientists believe that the current climate crisis could become a climate catastrophe that would be impossible for humans to mitigate. The trick to preventing such a catastrophe is to keep the permafrost frozen, and the Zimovs have a plan for doing that.

This unique, carbon-rich soil, called *yedoma*, sits in a basin as big as Texas and carries ten to thirty times more carbon than non-permafrost mineral soils. The National Academy of Sciences estimates that between 1,700 and 1,850 billion tons of carbon is trapped in the permafrost, which is more carbon than is present in the Earth's atmosphere and all vegetation combined. But Max Holmes, who studies the permafrost's carbon contents, tells me the deposits are a little closer to 1,500 billion tons. Today, as the global climate is warming, the permafrost will thaw, and we can expect a lot of that carbon to leak into the atmosphere. "There's an incredible amount of carbon locked up in permafrost in the Arctic, and a whole bunch of it's going to be vulnerable over the coming decades," Holmes says. Some of it is already thawing, and he thinks the speed of that thawing is likely to increase.

The carbon from plants and animals that died thousands of years ago is not dangerous in itself, but its decomposition is. When carbon-rich organics are exposed to the elements, bacteria chew away at the stuff, producing either carbon dioxide or methane, which gets released into the air. Those are both greenhouse gases that have contributed significantly to global warming and the climate predicament we already find ourselves in. Depending on the concentration of oxygen in the environment—which is, to some degree, a function of the amount of moisture there is—the carbon will be transformed into one or the other of the two greenhouse gases. Under conditions of little or no oxygen, carbon tends to be transformed into methane. "You're losing both CO_2 and methane from that environment," Holmes says. "But methane is the far more powerful greenhouse gas." That's the bad news. The good news is that methane doesn't last as long in the atmosphere as CO_2 does. "So in the short term—the short term being decades or even a century—you'd much rather have it go up as carbon dioxide than methane. Methane would cause a lot more warming."

There is no solid reading on how much carbon has already been released from yedoma soils (though Holmes thinks it is probably not yet that big of a number). Instead, scientists like Holmes have to work with what they do know, which is that there is about twice as much carbon trapped in the permafrost as what is already in the atmosphere. "So that's a whole bunch of carbon, y'know," Holmes says. "That's not all going to go from the ground to the atmosphere in a year or a decade or even a hundred years, but it's not unreasonable to think that we could start losing two or three petagrams per year." According to him, humans currently put 9 to 10 billion tons of carbon into the atmosphere annually. So our anthropogenic emissions are very large and the situation may get much worse, considering that 1 petagram equals 1 trillion kilograms (over 1 billion tons). Holmes is trying to figure out how likely it is that a significant amount of carbon from the permafrost is going to start heading into the atmosphere and, importantly, at what rate.

"Are you worried about what might be happening?" I ask. "Yes," Holmes says, "it is something I worry about. I wouldn't use the term 'doomsday,' but I certainly imagine carbon from permafrost making it a whole lot harder for us to get a handle on stuff. I mean, right now, humans are putting 9 billion tons of carbon into the atmosphere each year, mainly from fossil fuel combustion, secondarily from tropical deforestation. Both of those things, if we really put our mind to it, we could stop or slow. But with the permafrost, once it starts thawing, it's not directly under our control, and I think that's quite likely to start happening."

Sergey Zimov, who has been publishing scientific articles about carbon stores in the Arctic since the 1990s, is passionate about slowing the thaw. If something isn't done about it soon, the thaw could eventually warm the climate twice as much as burning fossil fuels already has. Zimov believes that the best way to keep the carbon locked up in the permafrost is to restore the ecosystem to what it was like when the woolly mammoths roamed there during the Pleistocene. At that time, the area was covered with rich grasses, which reflected light from the sun and did not absorb much heat. As the animals grazed all day, they trampled other, darker light-absorbing plants and carved holes in the snow with the force of their feet. A three-foot layer of snow on the ground is like a three-foot-thick insulation blanket.

If the outside temperature is -40 degrees Celsius (-40 degrees Fahrenheit), then it might only be -5 or -10 (23 or 14) beneath the white layer. But when millions of feet are punching holes in the snow in search of a blade of grass to eat in the winter months, that insulating blanket is broken and cold air can reach the ground. Therefore, Zimov says, Pleistocene megafauna like woolly mammoths acted like mobile ventilating systems. The air circulation they caused as they walked around kept things cooler than they would have been otherwise. They were also likely to have been landscape modifiers, as elephants are today—disturbing the ecosystem by knocking over trees that stood in their way as they browsed the available vegetation.

Herbivores that currently live in Pleistocene Park—and that heavily graze in certain parts of it—have allowed Sergey Zimov to monitor the effects grazing has on the surface temperature of the permafrost. In an article called "Mammoth Steppe: A High Productivity Phenomenon," he and his colleagues write that the winter temperature of the soil's surface in an area that had been subjected to long-term heavy grazing by herbivores was 15 to 20 degrees Celsius (27 to 36 degrees Fahrenheit) *colder* than in areas of the park where there had been no grazing at all. Furthermore, by fertilizing the soil with their dung, grazing animals kick-start nutrient cycling in the ground, enabling more grass to grow. The grazing sprees they indulge in during the shorter summer months have a similar effect of regenerating the vegetation by ripping away dead grasses, allowing new shoots to sprout. So if millions of animals were brought back to his park now, Zimov predicts, rich and light grasses would rebound as the dominant vegetation in place of the darker shrubs and trees that are there today, which do not reflect as much light and absorb more heat.

In 1996, the Zimovs put some horses and moose into half a square kilometer (124 acres) of the park and built a fence around it. In 2004, they introduced the first reindeer into

16 square kilometers (6 square miles) of land, which they enclosed with a much bigger fence. Then came the muskox and bison and wapiti. If the Zimovs' curated ecosystem can sustain itself without human assistance, they'd like to expand it to as much land as they possibly can. The hope is that in a few years' time, they will release more animals into the 500,000 square kilometers (193,000 square miles) of habitat that lies beyond the fence, between the Indigirka and Kolyma rivers, and more if they're able.

I ask Holmes what he thinks about Zimov's seemingly quixotic plan and am pleased to hear that he thinks it is solid. The big question for him is whether you can actually scale it up to a level that will make a difference on a planetary scale.

Going from several square miles to the size of Siberia and beyond is no easy feat, but the Zimovs are made of something special. Where others would have surrendered years ago, this family is fearless in the face of the challenge. "I'm living a rather exciting life, let's say, but sometimes, I'm afraid I can die very early doing this," Zimov's son Nikita tells me one winter night as I chat with him online from the comfort of my bedroom in Copenhagen. As his father was approaching the end of his sixth decade on Earth, he was spending his time setting up another rewilding project, located just south of Moscow, called Wild Field. Nikita is now the one responsible for keeping Pleistocene Park alive while his dad gets his new project off the ground. Reflecting on his inherited responsibilities, Nikita says, "I am rather happy with my life. I have a good family; I have three kids. I'm less passionate than my dad, definitely, but, let's say, it's a good idea and I just want to get it done."

When his dad was younger, he loved to hunt, but the tundra ecosystem didn't prove very useful for that. The tundra in their region is extremely difficult to walk through, which can make quiet sleuthing nearly impossible. Nikita had a horrible experience at the age of fourteen, when he got lost and had to spend two days all alone on the tundra, trying to find his way through the dark, wet, and cold shrubs back to civilization. Now, scarred from the experience, he says, "I want to burn that place to hell, like, just burn it with napalm because it is so impossible to walk through." If the unproductive tundra had instead been grassy steppe, covered by short grasses as far as the eye can see, it would have been much less harrowing to get lost in and certainly better to hunt in. "That's just so much nicer from my dad's perspective... so I think when he first came to this Pleistocene Park idea, it was actually his personal needs he wanted to satisfy."

The Zimovs and Pleistocene Park have been written about extensively before. "You know, we're rather spoiled with media attention," Nikita tells me, "and there were so many things written about Pleistocene Park which weren't really true that, at some point, I just stopped reading and caring. So write whatever you like." Caught a bit off guard, I ask, "What untrue things have people written about you?" "I don't know—that we are crazy. That's definitely not true. You know, there's Greenpeace, there are crazy scientists, there are fanatics, but we are not like that. We are very pragmatic and we can provide scientific support for people. I'm not very romantic, so when people say that we are crazy scientists in Siberia trying to create... trying to... The point is—we are not crazy!"

And I don't think they are either. I'm actually quite taken with their idea, even if this plan to bring live animals back started with an interest in killing animals. Morally speaking, Nikita sees Pleistocene Park as their way of helping humanity repay nature for the damage we've caused. Many scientists believe that humans hunted the megafauna out of the mammoth steppe at the end of the Pleistocene and consequently diminished the ecosystem's productivity, and he'd like to restore that gaping hole we left in nature with his work. He firmly believes that "humans came and broke the ecosystem, which was more efficient than what we have now... Also, becoming famous for doing something good for the planet isn't a bad idea."

However, the idea that humans overhunted the megafauna of the late Pleistocene—known as the "Overkill Hypothesis"— which led to their extinction and "broke the ecosystem," is not particularly well accepted by some scientists. The evidence for this phenomenon seems slim to a considerable number of archaeologists, paleontologists, and paleogeneticists, but it is disproportionally reported in the popular press as settled science. Although the debate is active and ongoing, some individuals, of whom there are a few in de-extinction, have concocted moral arguments for their work based on the side they believe in.

So far it's proven difficult to get the animals the Zimovs need into Russia to repopulate Pleistocene Park. They are often extremely expensive and can easily get caught in an international red-tape limbo. In the past, the Zimovs had to rely on the government to facilitate the transport of their herds that were sent by sea or flown in from areas outside of Europe, though they are now self-reliant. At one point, the Canadian government donated around thirty bison to their project. After spending a lot of time working with the local bureaucrats to get the bison into the park, Sergey was suddenly told that the bison that had already been admitted into the country had been assigned a new home in the region, which was not in their park.

"Ever since then we have had a very bad relationship with those guys," Nikita says. Now they manage all of their animal translocations privately and pay for them through international research grants and from the fees they collect from the scientists they host at the research station throughout the year. The Zimovs own another herd of twenty American plains bison, but they haven't been able to get them into Siberia. They got held up at the French border, and at the time we first spoke, had already been living there in limbo for a full year. Before the Zimovs could ship them to Russia, a virus known as Schmallenberg had spread through cattle populations, and Russia's borders had been closed for animal transport. There's a small post on Pleistocene Park's website that reads, "Does anyone have a desire to buy a herd of bison in France?"

Years ago, Nikita experienced one of his life's grandest challenges when he tried to wrangle some wapiti into the park. "I was driving from Moscow through the Altai Mountains all the way to Chersky in a truck where we had six wapiti, and it took me twenty-five days. It was a nightmare. It was absolutely off road, it was just a path, and I was driving through broken ice and everything was flooding." He'd bought a brand-new truck for the journey; after the expedition, it was unusable trash. Imagine spending twenty-five days with six wild animals in a truck, driving through frozen rivers with no road in sight, when the whole point is that you have to somehow keep them alive so they can make it to a faraway land. Amazingly, Nikita succeeded, and none of the animals died. "Yes, well, I got them there fine, but within the next six to nine months, some escaped, some died, and I think those who escaped also died... Certain animals turn out to not be the best idea."

Nikita is full of stories like this, like the time he went to get muskox from Wrangel Island, an extremely remote place in the Arctic Ocean between the Chukchi and East Siberian seas, where the last refugial populations of mammoths are said to have lived. Normally people get there by helicopter, but for this trip Nikita decided to go with some others by boat. They anticipated a week-long round trip but spent the first four days just drifting in the ice looking for a spot to dock, which they drastically overshot, ending up near Alaska. At long last they found the island and the nature reserve workers who live there. It had been agreed beforehand that the locals would catch seven muskox babies for Nikita to take back, and he was relieved to see they'd followed through. But on the first night after their arrival, when everybody was sound asleep, a polar bear broke the fence surrounding the corral, killed one of the muskox babies, and let the other six escape. Remembering it now, Nikita furrows his brow and I hear him sigh loudly through my speakers.

"Then we had to spend ten days looking for new babies because we couldn't find the original ones," he says. They'd made the trip in September, a month of constant fog and wind and rain, which they arduously withstood until they found six more babies to bring back with them. The problem now was that they were all male. They wouldn't be worth anything to the park past one generation. Nevertheless, the group arranged to bring them back by sea. Then, the day before they were going to be loaded onto the boat, the babies escaped again. "So we actually tried to catch them, running through this island, and at some point I was lost in the fog. I suddenly realized that there was no one around me, and I was just running on this island with a very high polar bear population all by myself and I had absolutely nothing with me. And I remembered a big billboard I once saw which explained how you should act when you encounter a polar bear. They were saying you should use a stick, and I thought it was funny, so I just ran faster and faster!" He leans forward, looking straight into the webcam as though peering into my eyes, and says, "It's a horrible island, I tell you."

In early 2000, Sergey bought a civilian version of a Soviet military crawler that he intended to use as a mammoth-mimicking machine. "So mammoths, one of their main duties was that they allowed the expansion of the steppe ecosystem into the forest by knocking down tree stumps," Nikita says, "and that's what this tank can do." Luke Griswold-Tergis, an American documentary maker working on a film about the Zimovs, has been to

Pleistocene Park and seen the tank's tracks circling tree stumps the pair had knocked down with it. He has talked to people who rode along with Sergey when he was "playing mammoth." They say it was absolutely terrifying.

For years, people have been asking the Zimovs if they would welcome cloned or recreated mammoths in Pleistocene Park. In 2014, Nikita traveled to San Francisco and Griswold-Tergis took him to meet the brains behind Revive & Restore, who presented him with a deal: ecological research from Pleistocene Park in exchange for a revived mammoth one day. Nikita understands why people involved with the woolly mammoth's de-extinction would want to partner with him and his dad. "They need scientific support for this," he says, "and our Pleistocene Park idea is the most developed in terms of justifying why we need to bring back a mammoth."

And that's not just his interpretation—it's become part of the de-extinction project pitch. As George Church (who leads the Woolly Mammoth Revival project that Revive & Restore supports) writes in *Scientific American*, "A few dozen changes to the genome of a modern elephant—to give it subcutaneous fat, woolly hair and sebaceous glands—might suffice to create a variation that is functionally similar to the mammoth. Returning this keystone species to the tundras could stave off some effects of warming." Church tells me that in conversations he's had with the Zimovs' team, the target number of unextinct mammoths suggested to make a beneficial difference to the ecosystem came up around 80,000. The thing is, though the mammoths were important for the steppe ecosystem that the Zimovs are trying to restore, neither father nor son particularly cares much about recreating them. As long as they have enough animals affecting the landscape—the right amount of bioabundance—they won't fuss about which animals they are. "This might sound mean," Nikita tells me, "but it can have five legs or two heads or three horns or whatever; we are just counting them as kilograms per square kilometer." That the Zimovs never scour the permafrost for mammoth meat in the hopes of cloning the species one day is telling. The only reason they collect mammoth remains is to calculate how many roamed in the park during the Pleistocene. But they are not against having recognizable mammoths with two tusks and a thick shaggy coat back one day either.

"There is this theory that we should support the cloning of mammoths," Nikita says, "and they will be given to us. With the only mammoths in the world, there will be money coming for the fence and for keeping the staff who will take care of that, so it will go to the park's development. It benefits us, same as them, so I don't mind." But Sergey, whom Nikita describes as a bit of a showman, handles the question differently. At a de-extinction workshop that Revive & Restore organized in 2012, he told the group that he would warmly invite revived cave bears, cave hyenas, reindeer, woolly rhinos, and woolly mammoths to Pleistocene Park.

Nikita says their official stance on the matter is that, in general, they support mammoth de-extinction efforts. They wish the revivalists luck and hope that they will eventually succeed. They believe, however, that they are trying to do something much grander and much more difficult than de-extinction. Instead of trying to clone a single extinct species, they are trying to bring an entire extinct ecosystem back to life. And they are doing it on their own dime.

A Mammoth Task

ON A BLUSTERY winter day in 2014, I took a bus from Toronto to Hamilton to meet with Hendrik Poinar—the paleogeneticist whose father's ideas informed the story of *Jurassic Park*—at

McMaster University. Poinar has been thinking about de-extinction for far longer than anyone else I was able to find, and not only as a scientist. His thinking started in his early childhood, when he'd do comedy routines with his dad about resurrecting extinct species from amber-trapped DNA, and it continued into his youth as a budding young biologist and consultant on the set of the first *Jurassic Park* film. If there were such a thing as a de-extinction aristocracy, Hendrik Poinar would be part of it.

Poinar's office is a little like a toy store. A tiny mammoth flag scrawled in Cyrillic that he found in Siberia hangs from his door-knob, and the bookshelves are lined with tiny plastic figurines of hose-nosed animals of all types. Some are large, some are small; some lift their trunks triumphantly, while others drag them on the ground; some have tusks and others none. Tiny fragments of real animal fossils are sprinkled between them. I sat down at the round table next to them so that we could chat, but something on the shelf kept catching the corner of my eye. When I turned to see what it was, I discovered, between two upright mammoths, a tiny, shaggy thing standing, bending an arm above its head. It was the figure of man, mid-lunge, about to thrust a spear right into the back of Poinar's chair. I looked down at my arm, which was bent as well, pointing a microphone up in Poinar's direction. In that moment, it was funny to think that I was using my technology to understand the effects that human technology had some ten to twelve thousand years ago.

It wasn't that long ago that the woolly mammoths roamed the Earth. It is a bit disorienting to imagine, but the last woolly mammoths and the Egyptians who built the pyramids lived at the same time. The bulk of the mammoth population crashed around 12,900 years ago, while the refugial island populations lived on until roughly 3,700 years ago. Research demonstrates that the population was responding well to the shifts that came with the glacial cycles of the last 200,000 years, colonizing across vast distances and breeding with other types of mammoths they ran into along the way. But then, about 12,900 years ago, their numbers drastically plummeted. Why didn't the majority of the population find a clever way to continue to adapt when that's what they had been doing all along?

When Hendrik Poinar was a kid, his dad used to take the family to France for summer research trips. And during those warm, sun-kissed months of his youth, he'd visit the caves where early humans drew mammoths, bison, and rhinos all over the interior walls. Those cave paintings called out to Poinar as a boy and gave him the sense that somehow the collective memories of those beasts were still roaming wild in his genes. He's been deeply fond of these creatures ever since. Although he has been contemplating de-extinction for most of his life, it is not an aspect of working with mammoth DNA that motivates any of his work. It's the evolutionary questions that interest Poinar the most.

Mammoths and elephants are part of a taxonomic group known as the proboscideans, which rose out of Africa around 60 million years ago. They are known for their incredible emotional intelligence and live in complex, matriarchal social structures; juveniles learn how to survive in the wild from their mothers and aunts. But the modern elephant is not descended from the mammoth. They each evolved from separate branches of the proboscidean family tree.

Although the northwestern tip of North America and the northeastern tip of Asia are separated by only 55 miles at the narrowest point today, things were different in the Pleistocene. Back then, much of the world's water was trapped in ice caps. Then the sea level dropped by 350 feet, and a strip of land that connects Siberia to Alaska emerged from beneath the water. This channel, called the Bering Land Bridge, was the migration motorway for mammoths, which walked back and forth between Asia and North America, sometimes interbreeding with distinct populations on either side of the land bridge. While they were around, mammoths lived across enormous swaths of land on either side of that bridge, stomping their way through western Europe, China, Japan, and Siberia all the way over to the glacial ice sheets that ran through what is now Yukon and British Columbia, down into the northern parts of present-day California. In North America, Poinar says, mammoth remains have been found as far east as southwestern Ontario's Green Belt, Maine, and Nova Scotia. But the mammoths never made it to southern Asia or South America; nor did they ever reenter Africa, where their story began—though a dwarf mammoth managed to reach Crete, which is close.

We tend to think of all mammoths as the iconic woolly mammoth, but mammoths were much more diverse than that and, in some cases, far less shaggy. The woollies were only one of several species that descended from the earliest mammoth ancestor, which branched from the Asian elephants about 6 million years ago. The earliest mammoths were the direct descendants of tropical African proboscideans that settled in the lush forests of southern Europe long before any mammoths appeared. Then came the southern mammoth, *Mammuthus meridionalis*, which lived in Europe and central Asia. Other taxa arose, like *Mammuthus trogontherii*, also known as the steppe mammoth because of the grasslands it inhabited across Eurasia and North America. They were the first mammoths to show any sign of a shaggy coat of hair. Adding to the mix was *Mammuthus columbi*, the Columbian mammoth, which lived in North America. It was just about as large as the steppe mammoth but had a range that extended much farther south. Then, of course, there's the iconic woolly mammoth, *Mammuthus primigenius*, which rose to prominence around 50,000 years ago during the middle of the Ice Age and was especially well adapted for cool habitats with its

WOOLLY MAMMOTH

thick insulating skin, heavy coat, and small ears, which allowed less heat to escape. However, these are all features that are only rarely preserved in the fossil record, mainly in Siberian mummies. Usually, woolly mammoth remains aren't mummified and get identified by their teeth. Therefore it is not clear whether the woollies had thick fatty skin, a shaggy coat and tiny ears in all of the places where they occurred across their range, or only in the coldest parts.

There are other identifiable mammoth species to add to the list, like *Mammuthus jeffersonii* (Jeffersonian mammoth) and *Mammuthus exilis* (pygmy mammoth). But Chris Widga, who describes himself as a "paleontologist/archaeologist/paleoecologist" at East Tennessee State University, tells me, "Mammoth taxonomy is a mess!" Distinguishing one mammoth species from the next by way of genetic analysis has made Widga's own work and that of his colleagues both simpler and more complex, because a genetically defined species may be different from a morphologically defined species. This conundrum is nothing new in taxonomy, but it makes spelling out all the mammoths a muddy task.

The woolly mammoth is believed to have existed in very large numbers and likely first arose in Siberia. Although cold, their environment was rich with grasses, a food they ate—and shared—with woolly rhinos, cave bears, reindeer, muskox, horses, bison, and giant deer. Compared with the other mammoths, their backside sloped more steeply, their tusks were more imposing, and their coat is thought to have been more unruly. Imagine an Asian elephant—the type you might see in a zoo—shrink it down just a bit, throw some hair on it, and curve its tusks: that's what the woolly looked like. It was also on the smaller side as mammoths go, except for those with evolved island dwarfism.

Dwarfism, in biology, is an evolutionary process that happens when animals get stranded on small islands that have scarce food supplies and so are at an advantage if they can survive without much sustenance. As a result, over generations, they end up being smaller. During the Pleistocene, the sea levels were so low that some of the California Channel Islands—where dwarf mammoths have been found—were connected to each other above water (though the sea level was not low enough to connect them to the mainland). So logic says that the ancestors of the dwarf mammoths that lived on those islands must have swum there. Though not technically dwarves, significantly small mammoths have also been found on Wrangel Island, where Nikita Zimov once ran for his life from possible polar bears.

Mummified mammoth remains and intestinal gut contents can tell scientists what the animals were eating, and isotopic signatures on tooth enamel can give clues about how far mammoths migrated over their lifetimes. But data from remains have been analyzed differently through time, and their interpretation has been complicated by the fact that the fields of ancient DNA, paleontology, and archaeology have been polarized for years between two competing hypotheses about what caused the mammoths to disappear.

On the one hand, there's the idea that climate shifted so drastically during the transition from the Pleistocene to the Holocene that they simply could not keep up. The warming climate caused forests to proliferate while the mammoths clung on to their grassy diets, but once the habitat changed so much that their food started to disappear, it was only a matter of time until the mammoths did too. On the other hand lies the thinking that inspired Poinar's work, which says that the large-game hunters that moved into North America at the end of the Pleistocene thought of the woollies as nothing more than walking steaks with fur coats.

Known as the Overkill Hypothesis (briefly mentioned earlier), this idea was popularized by an American scientist named Paul Martin. Martin argued that humans caused the widespread collapse of the large megafauna in Eurasia, North America, and South America—including mammoths—at the end of the Pleistocene. Considering that the gestation period for an elephant is twenty-two months, you can imagine that if you were to wipe out a group of female mammoths, you could annihilate an entire population very quickly. But as Poinar points out, there are problems with both theories. Even if the forest totally overtook the grasslands, we know that mammoths walked tremendous distances and could cross entire continents, no sweat. So if you were a mammoth in trouble, wouldn't you put those feet to use? As for the notion that hunters wiped out the mammoths, it is thought that the woollies were a breeding population of at least a million individuals. Was there really a group of marauding hunters so large that they could wipe all of them out?

To find the answers, Poinar searches for clues in their flesh and bones. "That's the most fun part of the job, really," he says.

"Traveling to the Yukon or Siberia and finding the samples." In many cases the samples have already been collected by other scientists and are stored in museums as frozen chunks of permafrost that still have preserved organic remains inside. In those cases, he will sample the mammoths directly in the museum, though he prefers to pull them out of the permafrost himself. "They're relatively easy to find in Siberia, about a dime a dozen, I would say. You really almost stumble over them." In Yukon Territory, where they're a bit scarcer, he can still find some lying along riverbeds wiggly enough to extract things from. When he's wrangled a mammoth chunk out of the wild, he scoops it up, takes it back into the lab, and then begins the arduous task of removing all of its mineral and chemical debris. That leaves an amazing pile of cellular muck—proteins, fats, and a ton of DNA. But the DNA is not pure, and turns out to be anywhere from 0 to 90 percent mammoth. In the worst cases it is 100 percent something else, which could come from anything imaginable—plants, rodents, bacteria, fungi, you name it. And there are cases where the contaminant DNA is very difficult to differentiate from the mammoth's, which makes his detective work trickier.

When that happens, the genome of a close living relative comes in handy as a template for how to put the fragments of DNA back together in the right order. The closest living relative to the ancient species is used because, compared with all other groups, these two species shared a common ancestor most recently in their evolutionary past. They should, therefore, have the smallest degree of genetic divergence between them. Ben Novak, a scientist who will reappear in these pages, once offered a metaphor to explain it that stuck with me: Much like assembling a puzzle from thousands of tiny pieces, scientists need to see the picture on the puzzle's box to know what image they are supposed to create with the pieces—and that's the template genome. The fewer genetic differences there are between the genomes being compared, the clearer the picture on the puzzle box will be. The Asian elephant is the woolly mammoth's closest living relative and so yields the best picture of the puzzle for the task. However, the problem with using a puzzle box from another species is that if there are any important genetic differences that make the extinct species different from the living species, the living relative's puzzle box will never include them. Scientists won't know how or where to map those important changes into the extinct animal's genome, and must accept that they might end up missing some important pieces.

But as long as the reference genome is assembled, the first step in piecing the extinct genome together is to sequence all of the tiny DNA fragments pulled out from fossilized remains, one chunk at a time, where each chunk might yield about one hundred genetic letters. Poinar explains: "In the old days we'd use PCR where we'd target stretches of DNA and amplify them up using what we call a genetic Xerox machine." PCR, or *polymerase chain reaction,* allows you to take a single copy of a gene and make millions of copies of it over a short period of time, which then get sequenced or read out. Today, scientists use high-throughput sequencing, which allows them to decipher the exact genetic letters in millions of DNA sequences within a few hours, and even faster methods are on the horizon. When Poinar was a grad student, one week's work would result in about five hundred of those letters. Today, one week's work for his grad students will easily result in several billion. Once the sequencing is done, the tiny DNA fragments they read out have to be placed back in the right order. Then the scientists must match them (like puzzle pieces) to the DNA fragments in the reference genome (on the box). Using statistical tests to analyze the matches, they assign stretches of woolly mammoth DNA to the Asian elephant genome this way, building it up piece by piece.

In 2015, Poinar and an international team of researchers led by Love Dalén, at the Swedish Museum of Natural History in Stockholm, published two complete high-quality woolly mammoth nuclear genomes. They had the best coverage of any genomes assembled of the species at the time, from two individual mammoths born 40,000 years apart. To do this, the team extracted DNA from a molar tooth of an approximately 4,300-year-old mammoth from Wrangel Island, one of the last survivors of the whole genus, and took the other sample from a juvenile Siberian specimen that was almost 45,000 years old. From their DNA, the researchers could demonstrate that the population was doing well across vast space and time and had survived two severe population bottlenecks (more on the bottleneck concept later). They found that the actual boom–bust of the mammoth population did not correlate as closely with climate change alone as many thought it did. Poinar already had a hunch that the reason for the disappearance of the mammoths was not one or the other of the two main competing theories but a delicate balance of the two. The researchers' findings reinforced Poinar's hunch, showing that the encroaching forest probably forced the mammoths into smaller and smaller areas, which they had to share with other megafauna, like bison, horses, and human hunters. But "ultimately," Poinar says, "I think, humans may have had the final blow."

"As for bringing mammoths back? Does your work fit into that?" I simply had to ask. He laughed, then stopped abruptly as all expression left his face, and said, "Well... sort of."

He knows that what he used to call bad sci-fi movie material has become a serious potential, in part thanks to research he has done. In an indirect way, he's turned the key in the ignition of mammoth de-extinction by providing high-quality, fully assembled woolly mammoth genomes to researchers worldwide. Now, anyone equipped with those woolly mammoth genomes can use them to engineer the Asian elephant genome to be woolly mammoth–like in the name of de-extinction. And that's not just hypothetical—such work is already underway.

Woolly Ways to Make a Mammoth

IN APRIL 1984, headlines raced around the world announcing that woolly mammoths were back. That year, MIT *Technology Review,* America's oldest journal of science and technology, reported that a Soviet-American scientific duo had teamed up to recreate the ancient extinct proboscidean through a process of hybridization with elephant sperm. The Soviet scientist, Dr. Sverbighooze Nikhiphorovitch Yasmilov, at the University of Irkutsk, had, it was said, recovered egg cells from a young woolly mammoth that had been frozen in Siberian permafrost, and had sent them to MIT's Dr. James Creak, who allegedly liked to take risks in his experiments. Creak, the article reported, had prepared some DNA extracts from elephant sperm, compared them with the nuclear DNA from the mammoth eggs he had been sent, and declared them compatible for hybridization. He then fertilized the eggs with Asian elephant sperm and implanted them in the wombs of several elephant surrogate mothers. According to the article, over sixty fertilization attempts failed before Creak was able to fertilize eight eggs with the sperm, only two of which were successful in bringing mammoth–elephant hybrids to term. But two were enough to announce the first mammoth de-extinction and the creation of a species Creak called "mammontelephas."

Phones rang off the hook, syndicated newspapers ran the story for months, and it appeared in over 350 publications in the U.S. alone before the journal revealed that their article was an April Fool's joke. Not only was it a hoax, but it had been crafted by a student, Diana ben-Aaron, who had cooked up the idea in an undergraduate writing class. The sham shone a mirror in the face of science journalism, which was so eager to sell hype that it didn't even question the validity of a story so outlandish released on April Fool's Day. But although the story contributed nothing to science, it did generate public discussion of what having hybrid mammoths in the world might mean. It also foreshadowed more sincere conversations about the prospect of mammoth cloning that the public would have twelve years later.

In 1996, Japanese scientists Kazufumi Goto and Akira Iritani set out with Russian partners on a quest to find frozen mammoth sperm or intact cells in Siberia, complete with all their nuclear DNA. Depending on what they could scrounge up, they would have liked to take a similar route to mammoth re-creation as ben-Aaron's hoax had laid out in MIT Technology Review. But today, over two decades later, and though others have joined the mission, nothing much has resulted from their search.

A newer effort to clone the mammoth from salvaged intact mammoth cells comes from South Korean scientist Woo-Suk Hwang via his cloning company, Sooam Biotech Research Foundation. Hwang is the once-disgraced researcher, expelled from Seoul National University, who fabricated results in one of the biggest cases of scientific fraud in modern history when he claimed to have cloned human embryos and generated stem cells from them.

When I open Sooam's website, I am greeted by a pop-up notification that reads, "When your dog has passed away, DO NOT place the cadaver inside the freezer. Then, patiently follow these steps: 1. Wrap the entire body with wet bathing towels. 2. Place it in the fridge (not the freezer) to keep it cool. *Please take note that you have approximately 5 days to successfully extract and secure live cells." A lot of Sooam's business comes through its cloning of people's dead pets for around $100,000 U.S. a pop. Animal cloning is a craft Sooam has honed well and has contributed a lot of knowledge to, but it doesn't want to be pigeonholed as a Fido factory alone. It's recently joined forces with China National GeneBank and Russia's North-Eastern Federal University to work on mammoth samples from Siberia in hopes of cloning a mammoth someday. Every few years a newly found mammoth cadaver is hailed as the best sample scientists have ever retrieved. And this gives the company confidence that eventually it will find the fully intact mammoth cell needed to get the cloning to work. As I write this, the search is still on.

The Woolly Mammoth Revival

THE FAILURE OF the more than two-decade-long search to find perfectly preserved mammoth cells (reproductive or not) signals little promise for cloning, and synthesizing all of the mammoths' chromosomes, with their right architecture, then wrapping them up in a nucleus without damaging them is still not feasible. However, one method remains that makes mammoth de-extinction possible and bolsters research that Revive & Restore is behind.

The Woolly Mammoth Revival is an ongoing research project overseen by George Church at Harvard's Medical School and Wyss Institute for Biologically Inspired Engineering. With no obvious application for human health or renewable energy—areas agencies want to fund—this work is just a side project of Church's lab. But Canadian postdoctoral fellow Bobby Dhadwar, who works with Church and his team, has been making enormous strides with the project. Dhadwar is fairly young, speaks moderately with a smile, and signs his emails as Sukhdeep—a name his grandmother gave him, though most people call him Bobby. The goal of their project is not to bring the woolly mammoth back to life but to engineer new mammoth–elephant hybrids that can repopulate the North American and Eurasian tundra and boreal forests, likely starting in Pleistocene Park. That's what is stated online at least, but when Dhadwar talks to me about the project, he offers a clarification: "When people hear about it, I think they get confused on the timescale. It's not like we are anywhere close to giving birth to a woolly mammoth."

The more immediate aim of the project is to take woolly mammoth genes that the team has studied from the woolly mammoth genome and introduce them into elephant cells to see what kind of an effect they have in elephant tissues. Kevin Campbell's lab at the University of Manitoba was the first to identify the functional relevance of genetic differences between mammoth genes and elephant genes, and found that specific genetic changes in the mammoth's DNA that coded for hemoglobin protein enabled it to carry oxygen in its blood at lower temperatures than elephants are able to. Campbell expressed the mammoth gene in bacteria, and then he compared its function to elephant hemoglobin in order to discover this. Now, by editing elephant cells with woolly mammoth–specific genetic sequences for traits like hair color, hair length, and the capacity for carrying oxygen in blood at freezing temperatures, Dhadwar is trying to see if he can reproduce specific woolly mammoth functions in elephant cells. Eventually, where elephants lack important traits that made the mammoths mammoth-like, he hopes they can edit those differences away and introduce the specific genetic sequences that would give them the traits they need to live like mammoths in the wild. Once the desired genetic edits are made to an elephant egg cell, the team hopes to implant it in a surrogate womb, where it will grow into a hybrid calf. "If we can move over just a few genes, we might not get a woolly mammoth, but at least eventually we might get a cold-tolerant elephant," he says, and to his mind, that's enough of a measure of success.

The woolly mammoth and Asian elephant are believed to have about 1.4 million specific genetic differences between them. That might sound like a lot, but when the entire genome is made up of several billion bases, as the woolly mammoth's is, it's merely pocket change. "When we look at the woolly mammoth and the Asian elephant, they are very similar," Dhadwar tells me. "The genes are usually identical and their differences are just at the level of SNPs. SNPs, commonly pronounced as "snips," are single nucleotide polymorphisms. A SNP indicates a genetic variant between paired chromosomes at the location of a single genetic base (A, C, T, or G), not at the level of the whole gene. So, to make an elephant genome 100 percent mammoth-like, researchers would need to make 1.4 million singular letter edits in the elephant genome with a precision gene-editing tool like CRISPR until both species' genomes are identical. But since they're only going for the "good enough" approach, that level of editing is not required.

To get started, Dhadwar and his team needed a cell to work in, because—to state the obvious—there are no woolly mammoth cells. "I know a lot of groups talk about going into the tundra and digging up a woolly mammoth and hopefully finding an intact cell with pristine DNA inside," he says, "but that simply does not exist. The DNA is too heavily degraded and cloning directly from a frozen mammoth is not practical." So instead of going the route of the Japanese or South Korean scientists, his group used elephant cells called *fibroblasts*—a type of cell that could be part of nearly any connective tissue. They got the fibroblasts from elephant caretakers, who regularly perform biopsies on their elephants to test if they are healthy. The fibroblasts arrive in the mail in the form of tissue scraps produced by ear-punch biopsies. At first the team was getting cells this way from African elephants, but with time they were able to obtain the more closely related Asian elephant cells from a placenta that a female birthed along with her calf at a zoo. Since the Asian elephant is more closely related to the woolly mammoth, these are the more optimal cells to incorporate mammoth-specific DNA edits into.

But those elephant cells need to be transformed into more useful cell types than ear-punch biopsies and placenta bits provide, for a couple of reasons. For one, Dhadwar's team is trying to test the effects of mammoth-specific genetic changes in elephant cell types *before* they engineer the full-blown beast. In order to study how the genetic variant that gave woolly mammoth blood cells their ability to bind oxygen at low temperatures will fare in elephants, for example, they have to first test that out in elephant blood cells. They don't want to test the expression of that genetic change in nondescript cell types like fibroblasts because those cells have little to do with how blood carries oxygen.

Second, after some generations fibroblasts stop dividing, thus restricting the time researchers have to work with them in the lab. Other types of cells, like stem cells, however, grow indefinitely, giving Dhadwar the time he needs to insert the mutations he wants into them. So an important step is to coerce the biopsied ear tissue back into an embryonic state in order to create from the elephant cells induced pluripotent stem cells, or iPSCs: adult cells that have been reprogrammed to functionally resemble pluripotent stem cells—undifferentiated cell types that have not yet been told exactly what distinct type of cell in the body they will turn into. Researchers can reestablish the state that cells are in before they are instructed to make different biological parts, like skin, an eye, or a liver. This is done by forcing the adult cells to express genes and other factors that define the properties of embryonic stem cells, which effectively sends them back in time to a moment when more was possible for their fate. Once they're pluripotent, scientists can decide what type of tissue they want to turn them into and coax them in that direction. Dhadwar's team is trying to get the adult Asian elephant cells they need through exactly these means.

But elephants are still around, so I wondered why the team would go to all the trouble of reprogramming stem cells when they could just get some from the animals themselves. The simple answer, Dhadwar tells me, is that they can't. As it turns out, zoos tend to give researchers access only to elephant tissues that they routinely extract from the elephants for health checkups. Stem cells make up only a very small population of cells in the body and are therefore very difficult to extract. Blood samples are often the easiest way to get a good yield of them, but there is only so much blood that researchers are allowed to draw. And when experimental stem cells are successfully harvested from the blood of other animals, like mice, the animals are pre-injected with a species-specific drug that helps mobilize the stem cells in their blood, making them easier to extract. But zoos are not willing to inject their elephants for other people's research. So Dhadwar's team must get their stem cells the hard way.

While the cells were being reprogrammed into iPSCs, Dhadwar set out to make some mammoth-specific genetic changes in the readily available fibroblasts in order to test that their gene-editing tools are working properly. It is the genetic differences unique to woolly mammoths, such as those that gave rise to their thick shaggy manes and layer of fat under the skin, that Dhadwar takes note of now. At first, while the researchers were waiting for their iPSCs to get made, they successfully edited fourteen genetic changes, in the form of SNPs, directly into the African elephant fibroblasts they had kicking around in the lab. "We edited these fourteen SNPs very easily," Dhadwar tells me, "and if we can do that in fibroblasts, we can do that in stem cells as well."

Some of those SNPs were woolly mammoth–specific, such as the genetic change responsible for giving woolly mammoths their hair color and ability to bind oxygen at lower temperatures than elephants can. However, some were "best guesses," Dhadwar says, and actually came from dogs, cats, and even humans with werewolf syndrome—a condition where people grow an abnormal amount of hair all over their body, sometimes covering their entire face. Dhadwar and his team chose genetic changes from these other species because they thought they were responsible for the thick, shaggy hair that these species share, which woolly mammoths also sported. But over time they were able to learn a lot more about mammoth-specific genetic changes by comparing a variety of more recently published woolly mammoth genomes. As a result, they no longer need to reach into the genetic libraries of cats, dogs, and humans to test the effects of gene editing in elephant cells.

When the researchers were finally able to obtain Asian elephant cells, the reprogramming of the iPSCs was still not complete. But because they finally had the right species' cells to work with, they could perform even more powerful gene-editing experiments in them. They are now trying to coax the Asian elephant cells into iPSCs. And when we last spoke, Dhadwar told me that he has edited more than fifty woolly mammoth-specific genetic changes into these Asian elephant cells to test their efficiency, and that the cells are responding well. He is making the edits one at a time and soon will combine all of the fifty-plus edits in a single master cell. And as soon as the iPSCs arrive, he'll transfer all of those genetic changes into them, coaxing the iPSCs into specific Asian elephant cell types, like blood, fat, and hair cells.

If the mutations produce the intended results in the specific cell types that the researchers test, demonstrating that mammoth traits can be properly expressed in Asian elephant cells, the plan is to eventually edit an Asian elephant embryo's genome with all of the important genetic changes deemed necessary to give rise to an animal that is very mammoth-like. Here too, though, there are hurdles to surmount.

Female elephants ovulate every sixteen weeks, although they also skip years of ovulation as a result of pregnancy and lactation. With most animals it would be possible to use an ultrasound to locate the follicle where the egg is developing and harvest it from the ovary using an instrument that travels up the reproductive tract. But that's not so easy with elephants. First, just locating the elephant's vaginal opening is tricky because females have more than a few feet of canal, called a vestibule, between where their vulva sits facing the outside world and where their hymen is, way at the other end.

Another problem is that the elephant hymen remains intact even after the elephant has sexual intercourse, and though it ruptures when a female gives birth, it grows back after each pregnancy. Sperm can reach the egg only by passing through a very tiny aperture in the membrane. With artificial insemination tools, researchers have been able to get sperm this far and through the aperture, but to actually get an egg out, they must navigate an enormous depth on the other side of that membrane until they locate the egg-producing follicle—too deep for an ultrasound to visualize without some extra help. Laparoscopic surgery, in which operations are performed through small keyhole incisions, might be suggested in other species when the follicle is so hard to reach. The process typically requires that the animal's abdomen be inflated to allow for sufficient operating space and better visualization of the internal structures. But inflating the abdomen of an elephant could kill the animal, since elephants lack a pleural cavity (the space between the squishy membranes that surround the lungs and line the inner chest) that makes inflation harmless in other animals—the elephants' chest cavity could easily become overcompressed.

There are other ways that scientists might be able to access elephant eggs without hurting the females that produce them.

In the past, scientists have been able to take ovary tissue from one animal and insert it into the ovary of an individual from another species that then makes its eggs. The thinking is that by taking ovary tissue from a recently deceased elephant and transplanting it into the ovaries of another living species, they'll turn that live animal into an elephant egg farm. The late veterinary pathologist John Critser did this in the 1990s when he and his colleagues took frozen samples of ovarian tissue from African elephants that had lived in South Africa's Kruger National Park, thawed them, and then grafted them into mice ovaries. Consider that for a second—they modified mice so that they'd produce eggs from an animal thousands of times the size of a mouse! Their results showed that mature egg-making follicles were indeed developing, though the mice didn't manage to produce elephant eggs. If this experiment were improved upon, perhaps a procedure like this could create a source of elephant eggs for de-extinction pursuits.

I ask Dhadwar how he plans to get the elephant eggs they need, and he tells me, "There have been cloning technologies developed for other organisms, but translating and validating these methods to mammoths or elephants will take time. Speculating today about what method would be used does not account for continuous research currently being conducted in these fields. This project would not have been initiated without the development of CRISPR-Cas9 editing technology. I foresee similar breakthroughs being made in cloning and embryogenesis."

If they one day manage to insert all of the desired mammoth DNA into a fertilized egg cell, they will have to put it somewhere it can develop. But elephants are having a hard enough time as it is maintaining their populations, so Church and Dhadwar's team has tentatively rejected the idea of using real elephants as surrogates in its experiment. I agree that it would be better to free up today's remaining proboscidean uteruses to grow more elephants instead of new experimental necrofauna. For this reason, the researchers might look to develop artificial womb technology for the task. *Ectogenesis,* a term coined by British biologist J.B.S. Haldane in 1924, refers to the growth of an organism in some sort of vessel outside of the body. In the 1990s, Japanese researchers came up with a technique called *extrauterine fetal incubation,* in which they connected catheters to large blood vessels in goat fetuses' umbilical cords and fed oxygenated blood to the fetuses as they grew in tanks of amniotic fluid heated to a goat mother's body temperature. Dhadwar seems irked when I ask him what their artificial womb might look like. "For now we are just saying, let's take a look at a few genes, let's take a look at their functionalities, and then we will scale up from there." People always ask me how long it might be until we see a recreated mammoth in the wild. But if there's one thing I've learned from the scientists at the forefront of de-extinction, it is that we should stop asking them for a realistic timeframe. At least, I'm not holding my breath.

A Genetic Scientist against Mammoth "De-extinctification"

IF DHADWAR AND his team ever recreate a mammoth-elephant hybrid, they will still not have met their goal unless the animal can withstand cold Arctic temperatures. But another scientist researching the woolly mammoths' cold tolerance leaves that work in the lab and refuses to take it into the field. Vincent Lynch is a professor in the Department of Human Genetics at the University of Chicago who, like Poinar, has sequenced the woolly mammoth genome. He has a thick head of shaggy brown hair that is not totally unlike that of the charismatic species under study. Although most of Lynch's research is focused on trying to understand how pregnancy evolved, he has identified a set of genome analysis tools to answer a broad range of questions about pregnancy, and he decided to test them on the woolly mammoth genome. He is fascinated by mammoths but says that no matter what, he will not be involved in any "de-extinctification." "We will be resurrecting mammoth proteins and testing them in the lab, but that's where it stops," he says. Nonetheless, he understands that by publishing the woolly mammoth genome, he, like Poinar, has provided woolly mammoth revivalists with the knowledge they need to do their work.

Lynch's critique of mammoth de-extinction is many-sided. "How many of the woolly mammoth genetic mutations do you have to incorporate into an Asian elephant genome, and how many Asian elephant–specific changes do you have to edit away before you can call your edited Asian elephant a mammoth? Is it enough to produce one hundred changes? I don't think so," he says. To Lynch's mind, for a true de-extinction to have occurred, the Woolly Mammoth Revival will have to make all of the approximately 1.4 million genetic changes where their genomes differ. But even then, he points out, it's not really a mammoth, but a manipulated Asian elephant. "What does it mean to be a thing? You can never make an Asian elephant a mammoth because you've done something to it. You will always have a transmutated elephant that will never be a mammoth. It's sort of a wishy-washy argument, but I stand by it just the same." This same argument, of authenticity, could be made about any of the de-extinction candidate species that would be gene edited.

Dhadwar, Church, and their supporters at Revive & Restore argue that the purpose of creating mammoth-like elephants is ecosystem recovery, not precisely honed extinct species recovery. To recap, because the grassy steppe was adapted to mammoths and fell out of whack after they disappeared, these researchers believe that bringing mammoth-like creatures back now could resurrect the tundra's productivity and even keep the permafrost there from rapidly thawing. But Lynch isn't buying it. "That argument is actually profoundly anti-evolutionary, because that environment has been adapting to their extinction. Ever since the mammoths started to become less abundant in the environment, the environment started adapting to their absence."

If mammoths are brought back now, he foresees that the tundra won't necessarily spring back to its former state. It could instead transform into something new altogether, which might have unpredictable downstream effects. The cases both for and against revived mammoths being able to restore old ecosystem dynamics are speculation at this point and have little evidence so far to support them. Someone will have to introduce a sizable herd of recreated mammoths up north to really find out what cold-tolerant elephants will do there. But is that a good enough reason to say that they *should* do this?

I ask George Church if he considers de-extinction to be an environmental necessity of our time, to which he says, "I am not sure of any necessity for anything. There's always other ways of doing things. But it seems like a very cost-effective and possibly the most cost-effective way" of helping slow the permafrost thaw. Although other types of grass-eating animals could be moved up north to run around and punch holes in the snow, he says, few would possess the same physical force as a mammoth—or as a cold-tolerant elephant—and therefore would be less effective at performing the ecosystem services that are sought after.

But will the 80,000 or so unextinct mammoths be able to be introduced into the Arctic fast enough to make the beneficial differences some are hoping the introduction will? What if climate change beats the science to the punch?

Once they're able to make the mammoth–elephant hybrid embryos, an accomplishment Church has estimated they'll arrive at in 2019, it will take nearly two years for that embryo to grow into a calf (if it gets implanted in a surrogate mother or successfully developed in an artificial womb), and another eight years for the resulting animal to reach sexual maturity. Given these timeframes, a lot of permafrost might in the interim thaw and release its carbon stores that will get turned into methane or carbon dioxide, exacerbating the issue that the recreated mammoths are meant to help fix. Reflecting on these issues, Helen Pilcher, a science journalist and comedian who wrote a book about de-extinction called *Bring Back the King*, writes that it would take well over a century to make a single viable herd, which is nowhere near enough mammoths to do the job. She boils it down to this: "In other words, we cannot look to mammoths to help solve global warming."

But Church says that in principle, 80,000 could be produced in parallel in eight years. He also brings my attention back to the potential considerable advantages and tells me, "There's all the benefits of revitalizing the conservation movement by the injection of new hope and new technologies. Conservation was worse than a zero sum game; it was a steadily declining game. It was like Napolean telling his forces, 'We're going to lose eventually and I'm going to be exiled but let's keep fighting,' which is like, 'We're going to lose all these species, but we're going to put it off as long as possible.' It's a very demoralizing attitude for which there is now an alternative." I have been captivated many times by that point while pondering the merits of de-extinction. I do believe that de-extinction could create a real sense of hope about the future of conservation for a lot of people. And that matters enormously! But it doesn't drown out the other questions I have about it.

One ethical argument against making a mammoth that I often think about has to do with their intelligence. Proboscideans have complex social structures, with matriarchal societies in which knowledge about how to survive in the wild is passed from mothers and aunts to babies. If we put a hybridized elephant embryo into an Asian elephant and it gives birth to a healthy calf with mammoth genes that needs to learn how to act like a mammoth, not an elephant, how will it do that? It's possible that at first scientists might create something that doesn't look much like either a mammoth *or* an elephant but something in between. What if the Asian elephant mother that is supposed to raise the hybrid calf doesn't recognize it as one of her own? Will she reject it? Then you have a baby mammoth-elephant that is essentially a social species living all by itself, which, as Lynch points out, sounds like a really sad existence. Some zoos are no longer keeping elephants at all, particularly solitary ones or those in small groups, because of the psychological stress it causes them. What kinds of psychological stress might we create for the first generation of recreated mammoths if they aren't properly accepted into the group? Will the surrogate's elephant ways influence the hybrid, making it something different from what researchers hoped it would be?

The environmental, social, and behavioral uncertainties are no tiny issues in mammoth de-extinction. Besides these, the technical hurdles to be overcome are still vast, though I have little doubt that they'll be solved in time. These are extremely smart people at the top of their fields who are working on making it all possible, in the context of wider biological-engineering research with lightning-fast rates of advancement. But overall, I think these obstacles are good for us. Although the proposed ecological justifications are interesting, I am not convinced that we should bring a hybridized woolly mammoth–like creature to life right now given that, as the Zimovs say themselves, the mammoth steppe ecosystem might be restorable without a re-created mammoth's participation—as long as there are enough heavy, hungry, and determined beasts running around to punch holes in the snow.

Mammoth re-creation is a fascinating scientific and technological prospect. Enormous progress will be made toward that end in the coming years, and much valuable knowledge will be generated from the attempt. But the feeling I'm left with is that we should be trying out as many *other* ways as we can to increase the mammoth steppe's productivity—with existing animals—before resorting to the romantic ideal of de-extinction. I know, what a downer I'm being! People protest all the time and cry, "But I want to see a woolly mammoth!" Well, the question I have for them is, at what cost?

If there were a way to ensure that the suffering of individual recreated mammoths would be minimized and controlled, and if significant ecological benefits of having mammoths running around instead of other means of ecological disturbance could be experimentally demonstrated with comprehensive models and laboratory work that show it could be done in an effective timespan, I'd be more comfortable vouching for some form of the mammoth's return. Among other things, there's a need for robust studies of how the current tundra ecosystem will react to the introduction of hundreds to thousands of hybrid mammoth-elephants, along with good legislation to protect the animals from being used for unecological purposes. What I am more interested to see is how the same technologies used in de-extinction might be used to help elephants do better in the wild, which Church and Dhadwar are also working toward, as chapter 7 will explore. In any case, we don't yet know what the outcomes will be, and along the way we'll learn a great deal. As Stewart Brand says, "This generation gets to rethink extinction, gets to rethink habitat loss and habitat restoration, and gets to ponder the role of biotechnology in protecting biodiversity. Welcome to a very interesting century." He's right—get comfortable; we'll be here for a while.

CHAPTER FIVE

CAN BILLIONS OF PASSENGER PIGEONS REBOUND, AND SHOULD THEY?

A Bunch of Birds in a Drawer

ON A FOGGY fall afternoon I find myself in the entrails of the Royal Ontario Museum in Toronto, getting a tour of the museum's archival collection. My guide is Mark Peck, a tall, sturdy man with short graying hair who is head of the museum's ornithology division. I traipse behind him through employees-only corridors, past shark skeletons and bone casts, into a room he manages that is replete with well-preserved dead birds. As a child, on family visits, I used to dream about getting lost in this very museum: after looking for me for hours and giving up hope, Security would unknowingly lock me in overnight and I'd see the exhibitions come alive. Magical as it is to finally see some of the museum's hidden specimens, their smells and their stillness leave no room to imagine that they contain even an ounce of vitality.

After introducing me to some of the technicians and letting me glance over the museum's collection of preserved owls, Peck starts off down a hall lined with big beige cabinets taller than he is. "I will open up the passenger pigeon cabinet for you," he says. He fumbles to get the doors open, but after three tugs they fan outward to reveal a set of drawers. The first one he pulls out uncovers row upon row of what the natural historian Joel Greenberg calls "mourning doves on steroids." Taken together, these drawers hold 132 passenger pigeons in all. The birds are dried, stuffed, and arranged side by side.

Peck, who has a perfect name for his career, pulls out a bird; dangling from a small piece of twine attached to its dinosaur-resembling foot is a tag. He reads out something scrawled on it in faded black cursive writing. "This one was collected in Hamilton in 1863. That's over 150 years old at this point," he says. He puts it back down and opens the next cabinet. "What do we have here? Oh, one and a half shelves of females. So, males are easier to identify. They tend to have a buffy orange breast associated with them, whereas the females tend to be more of a gray breast. On the back of the birds you get a nice gray on the males, and the females are a little more cryptically colored." Others have described male passenger pigeons as slate blue with a rich copper-colored breast mixed with a bit of iridescent purple. The females display much drabber versions of similar colors. Where the male's breast is copper, for example, the females exhibit a beige patch. The sexes are approximately the same size, though—about one and a half times the size of a mourning dove, or roughly the same size as a common rock pigeon. They're 15 to 18 inches long and weigh 10 to 12 ounces.

Most of the birds here are lying lifeless on their backs, with blank, glassy eyes staring out into some distant universe, but a few have been posed to look a little livelier. Of the ones that have been put into tableaux, some have their wings open, some have theirs partially closed, and two fancy-looking birds sit on branches with fully fanned tails. They've been sitting stiffly like this for decades. They seem lonely, and for the most part they are left alone, but these birds have been getting a lot of visitors recently. The year 2014 marked the centennial of their extinction, and passenger pigeon admirers, as well as scientists interested in the birds' DNA, have been flocking to see them. The Royal Ontario Museum holds the world's largest collection of preserved passenger pigeons; many of the samples used for the bird's de-extinction have come from this very same batch.

"The wonderful thing about passenger pigeons is that they are thought to have been one of the most common birds on Earth," Peck tells me. "They talk about a population of billions of birds darkening the skies for three days when the big flocks came in to feed on the acorns and beech mast. It was a remarkable bird in a lot of ways, and to think that a bird that was so common could go extinct in really fifty years, due to habitat loss and hunting, is hard to believe."

Genomic sequencing has shown that the passenger pigeon diverged from the last common ancestor it shared with its closest living relative—the band-tailed pigeon—anywhere between 9 and 16 million years ago. While I was writing an earlier draft of this chapter, that number was estimated to be between 18 and 22 million years, which roughly corresponds with the period when the Cascade Mountains formed, and it was thought that the mountains were likely the barrier between the species that made the band-tailed pigeon a western bird and the passenger pigeon an eastern bird. But a more recent publication shifted the numbers to the younger dates, which don't align as cleanly with the mountain formations. New ideas are now emerging about what might have caused the two species' evolutionary

PASSENGER PIGEON

differences. But both timespans describe an incredibly old lineage, and the passenger pigeon's lengthy existence as a species makes it all the more remarkable that it crashed in such a short period of time. In her book *Resurrection Science,* M.R. O'Connor—a journalist tracking biotechnological changes in the conservation landscape—makes the astonishing comment that passenger pigeons went extinct in about one-thousandth of a percent of the time they were alive on Earth.

Passenger pigeons ate a great many things, though it was observed that the fruits of hard-mast trees were their favorite food. However, that might have just been a byproduct of how abundant the trees that produce hard mast have been over the last 10,000 years. When the birds had their pick, they'd go straight for beechnuts, acorns, and at times chestnuts, which their flexible jaws and elastic throats allowed them to swallow whole. Some bird throats, like the passenger pigeon's, include a special pouch called a *crop*—a pocket-like area that can store food before the bird begins to digest it. The passenger pigeon's crop could hold up to half a cup of nourishment. Once it was full, it bulged out twice as wide as the bird's body. If a pigeon happened to be shot while it was in this stuffed state, records say, it would crash to the ground with the sound of a sack of marbles hitting the floor.

When the hard mast (acorns and so on) wasn't available, passenger pigeons would munch on locusts, various insect larvae, earthworms, and snails. Their devoted historian, Joel Greenberg, has found records of at least forty-two genera of wild plants they would eat, which isn't even counting the domestic plants—like buckwheat, rye, corn, hemp, and wheat—that they also picked away at.

But while passenger pigeons were busy looking for food, we were looking at them *as* food. "You could go into these huge breeding colonies, and you could smoke them out," Peck tells me. "You could shoot as many as twenty to twenty-five birds with a single shot. They would quickly have their feathers removed, be stuffed into barrels and salted." Greenberg puts the number even higher, noting that records say as many as ninety-nine birds could be shot at once. The young ones—the squabs—generally had a lot of fat on them, which made life a little easier when they first left the nest. Unfortunately for them, it also meant that they tasted better to us humans. Fatty squab was a common menu item in fine restaurants in New York, Chicago, and Toronto. Holding a stiffened bird in each hand, Peck looks at me with subtle mourning in his eyes and says, "There are stories about trains carrying nothing but passenger pigeons down to some of the bigger cities in barrels."

The first comprehensive study of the passenger pigeon's extinction was produced by an American chemical researcher named Arlie William Schorger and published in 1955. From the copious records Schorger collected, he was able to show that it was common opinion in the United States that fatty pigeons tasted best. The ones that fed on beechnuts were especially desirable for fine dining, since they were rich in tasty lipids. In order of preference, according to Schorger's records, human eaters seemed to like squabs first, birds raised in captivity second, and those that were hunted in the months of September and October third. Sometimes the birds were captured alive and moved into cages for fattening before they were eaten later in the winter.

People came up with a great many ways to preserve dead pigeons. Most often this was done with salt, but they were also covered with molten fat to create an airtight seal, then packed in casks for long-distance shipping. Other times, they were pickled in spiced apple cider and sealed away in jars. For meals, their breasts were smoked, dried, and jerked, adding some variety to their flavor. An 1857 guide intended to help settlers deal with day-to-day living in Canada describes a recipe for a savory pigeon pie: "To make a pot pie of them, line the bake-kettle with a good pie crust; lay out your birds, with a little butter on the breast of each, and a little pepper shaken over them, and pour in a teacup of water, do not fill your pan too full; lay in the crust, about half an inch thick, cover your lid with hot embers and put a few below. Keep your bake-kettle turned carefully, adding more hot coals to the top, till the crust is cooked."

But people didn't use passenger pigeons only for food. Schorger writes that the bird's gizzards were thought to dissolve gallstones, its blood was believed to help eye disorders, and its apparently anodyne dung was used to help headaches. Squab oil was good for soap stock, and squab feathers were a popular stuffing for comforters and pillows. According to records, at a certain point, no young woman in the vicinity of Saint-Jérôme, Quebec, ever married without a dowry that included a bed and pillow full of pigeon feathers; an old wives' tale declared that a person could never die if they slept on a bed of pigeon down.

The ways in which humans killed passenger pigeons could be incredibly creative and incredibly cruel. One depraved method involved gathering the squabs that hadn't learned how to fly by setting fire to the bark of the birch trees that their nests were balancing on. A Native chief named Simon Pokagon once described the scene: "These outlaws to all moral sense would touch a lighted match to the bark of the trees at the base, when with a flash, more like an explosion, the blast would reach every limb of the tree and while the affrightened young birds would leap simultaneously to the ground, the parent birds with plumage scorched, would rise high in air amid flame and smoke. I noticed that many of these squabs were so fat and clumsy they would burst open on striking the ground. Several thousand were obtained during the day by that cruel process." After the initial blast, men would appear wearing old torn clothing and burlap sacks wrapped around their heads and feet to protect them from the spray of pigeon feces that covered everything. The brutes wielded clubs to knock the remaining birds out of whatever nests remained. Etta Wilson wrote that "of the countless thousands of birds bruised, broken and fallen, a comparatively few could be salvaged yet wagon loads were being driven out in an almost unbroken procession, leaving the ground still covered with living, dying, dead and rotting birds. An inferno where the Pigeons had builded their Eden."

In the second half of the nineteenth century, once railway lines were in regular use, the pigeon market trade became an organized ecosystem all of its own. While researching the passenger pigeon's extinction, Schorger received a letter from a

Wisconsin man named Alvin McKnight who could offer some firsthand accounts of the pigeon trade. He described the nesting season as lasting for three weeks, long enough to allow 1,500 barrels of pigeons, packed tight with thirty to thirty-five birds each, to be shipped to New York City each day. His letter reads, "In packing for shipment, the ends of the wings were chopped off, but very few were ever plucked or drawn... My uncle Chas. Martin used to make around $500 a season trapping pigeons from one to two dollars per dozen... Sometimes when his catches were light, he would put them into crates alive, take them home and put them in a feeding pen and feed them about ten days. Then he would kill and ship them as plucked, stall-fed birds. For these he used to realize a fancy price, sometimes as much as $3.50 per dozen." That's roughly $105 today.

Trappers found strategies to optimize their return on the birds depending on the season. Dedicated hunters saw dollar signs flying high. That helps explain how the most populous species humans ever interacted with disappeared in less than half a century. Eventually, one too many trains left the station, and the billions of birds were gone. Then, just a few short years ago, Revive & Restore hatched a plan to get them back.

The Great Comeback

ON THE DAY we first met, I could immediately spot Ben Novak in the crowd. Twenty-six at the time, just like me, he was obviously younger than every other scientist in the room. It wasn't only his youth that made him conspicuous to me, though: it was also the nineteenth-century "gentleman scientist" vibe of his buttoned-up vest. Novak is tall and sometimes wears his brown hair in a heavy swoop that folds over to the left side of his face, exposing the side of his head that is shaved. That day, however, his hair was parted down the middle. He figured he should go for something more conventional than his usual asymmetrical style in order to come across more professionally, even though he hated how it looked. We chatted in the lounge outside of the National Geographic Society's auditorium moments before he took the stage to tell the world about his plan to bring the passenger pigeon back to life.

Novak is the lead scientist of Revive & Restore's flagship de-extinction project, officially called the Great Passenger Pigeon Comeback. By now I've met with Novak multiple times— in person, in radio booths connecting Santa Cruz and the CBC broadcasting center in Toronto, and online—to keep up with the great complexity of his passenger pigeon pursuits. As of my writing this, he has been on Revive & Restore's payroll for over four years and has recently published his first scientific paper. Despite not being a more senior researcher, he is standing steadfast at the center of a scientific mission that could dramatically advance a field even the most accomplished scientists hardly know anything about. If passion is a valid stand-in for a publication record, he is no doubt the best man for the job. When Novak talks about passenger pigeons, he radiates an inspired sense of boundless enthusiasm and old-world romanticism that for some might be overly zealous. But I've always appreciated how he brings a dash of shameless gusto to research that others often eschew in the name of "serious science."

"You know the funny thing is that there's a world of people that get bit by the pigeon bug and they go crazy," he tells me. "Pigeon fanciers and others who just fall in love with them— there's a select group, and when they discover this story of the passenger pigeon, they become very passionate, almost entranced or enthralled." Novak counts himself among them and is always animated when he talks about the bird. He came to the passenger pigeon through an already developed love of science and the idea of studying extinct species. He first heard about the bird in junior high. Back then, he read an article in *Audubon* questioning whether science could ever bring the Tasmanian tiger back to life. Around the same time, he read another article, in *National Geographic,* that introduced him to the concept of human-caused extinctions, which he'd never contemplated before. "I mean, we get endangered species thrown at us from a young age, you know," he says. " 'Save the panda' and 'Stop killing elephants for ivory' and all of that. But we rarely got to hear about animals that we've *lost.* Socially, we are ashamed of it, but this *National Geographic* article put it directly out there."

It really shook him. He put two and two together, and with that, his career goal was set at the age of thirteen. The future of conservation was, he felt, going to be about restoring the gaps in nature that humans had caused, and he prepared a project for his school science fair about the technological problems one would have to overcome to resurrect an extinct species. "I chose the dodo bird because I loved birds a lot, and the dodo bird is the icon of extinction. 'Dead as a dodo,' as they say. I learned through that that the dodo is actually a giant extinct pigeon, and learning that about the dodo gave me the pigeon bug."

From dodos to passenger pigeons, his curiosity soared, on an emotional level as much as an empirical one. While exploring his new fondness for pigeons, he came across a beautiful photograph of a passenger pigeon in a book. It struck him as an exceptionally regal specimen. Then he read the caption: the bird in the photo was extinct. A single flock of more than a billion passenger pigeons, the caption added, took over a dozen hours to fly over any one spot and blocked out the sun as it soared. It felt like a thing of fantasy, and Novak was immediately hooked.

"Everyone falls in love at some point in time with something that's imagined," he says, "whether that's unicorns or great armies of Orcs fighting battles of good and evil. And here I was, a big Tolkien fan at that age, reading about flocks of billions!" The immensity of the flocks meant that the passenger pigeon made sense to him only as myth. But over time, he felt what he calls an "extra dimension" set in. No matter how magical they appeared in his mind, he knew that once upon a time, passenger pigeons had been real birds that real people saw with their own two eyes and killed with their own two hands. It gripped him. There was something about the tragic fact that he would never get to experience these enchanting birds up close in real life, that they were at once real but also sort of supernatural. He couldn't let it go.

When he got to university, he began to design what was intended to be a population genetics project about the bird, but he needed to get his hands on some tissue samples. That proved to be very difficult. He spent a few years requesting tissue from institutions that he knew had their own passenger pigeons. He dutifully persevered through their multiple rejections until finally, a year and a half before our first interview in 2013, Chicago's Field Museum agreed to give him a sample from its collection.

By that time Novak was working in Hendrik Poinar's ancient DNA lab at McMaster University, where he was studying mastodon fossils. He was able to use the resources in Poinar's lab to do the initial processing, then sent the samples away for sequencing at the University of Toronto, which he paid for with funds that he received from friends and family. Before he got any of the genetic sequences back, he had only visualized the bird's genes in pictures he had painted on canvases filled with its plumage color palette.

One day, while his population genetics investigation was still underway, Novak got an email from Joel Greenberg, the natural historian of the passenger pigeon, telling him about a meeting that was going to take place about the potential de-extinction of the bird. When Novak discovered George Church among the notable scientists from the bleeding edge of biotech on the list of attendees, he emailed him to say that he was working on passenger pigeons and would love to be part of the de-extinction project. Church forwarded that email to Revive & Restore, and with that, Novak's childhood dreams became real. He was hired by Phelan and Brand as the lead scientist for their first official de-extinction endeavor, spent a month in Church's lab in Boston, and eventually relocated to Santa Cruz to work in Beth Shapiro's paleogenomics lab at the University of California. Novak's research on the Great Passenger Pigeon Comeback found a home, and he no longer has to beg museums to send him tissue samples for his research. He just gets down to work with the best of them.

Laying Eggs for the Lab

BETH SHAPIRO IS a sprightly and direct speaker. With her bold brown eyes, shiny brown hair, and polished cadence, she seems television-ready when she lectures. She originally wanted to be a broadcaster—as you can tell by the lively and clear way she speaks to the public about her scientific work. But one of the first things she said in our interview was not at all clear to me: she told me that she was definitely not interested in bringing the passenger pigeon back to life. "We're a genomics lab—we are interested in understanding genomes and evolution, but that's as far as my research with this particular project is going to go," she said. That was a puzzling thing to hear from the principal investigator of the lab that hosts Novak's research on that very bird's potential return.

In 2001, Shapiro was involved in a project in which, together with some colleagues, she tried to figure out what other types of birds the dodo was most closely related to. Many incongruent hypotheses had been made, and she wanted to test them out. To do that, she and her team targeted short fragments of dodo DNA from the bird's mitochondria. "They can tell us about the amount of diversity in a population, the rate of diversity through time, and they can take us back past things like population bottlenecks that might have happened," she tells me. Since there are several copies of mitochondrial genomes in a cell, mitochondrial DNA provides useful redundancy to researchers, especially considering that DNA starts to degrade as soon as an animal dies. In 2001, the technology was such that paleogeneticists like Shapiro could amplify very short fragments of mitochondrial DNA from the bones, hair, and teeth of many extinct species.

Mitochondrial genomes help solve mysteries about how populations lived, but they also provide information about how species are genetically related to each other. By using mitochondrial clues, Shapiro's team was the first to discover that the dodo was part of an evolutionary group of morphologically diverse pigeons. Considering the dodo's genetic connection to Novak's favorite bird, Shapiro became interested in the strangeness of the passenger pigeon's abundance and its lightning-fast crash. Sure, they were hunted for food, but why didn't at least some of these birds, when there were billions of them, manage to adapt, evolve, and survive? Shapiro says that "it is a tremendously weird mystery, and a mystery that is worth solving."

This she set out to do after her dodo project was completed. The thing was, she needed to first assemble the passenger pigeon genome. Novak needed that for his de-extinction project too, so they worked together. And that's why it isn't confusing after all that Novak was sent to work in Shapiro's lab even though she was not interested in resurrecting the bird: they shared a scientific goal. First on their to-do list? The passenger pigeon genome. For that, Shapiro had sequenced some passenger pigeon DNA. Luckily, Novak had extra passenger pigeon DNA to donate to the cause.

What is a day in the life of an extinct pigeon resurrector like? For starters, Novak says he does a lot of wet-lab work. Roughly speaking, there are two main realms of working with genetics: the wet lab and the dry lab. The former is largely hands on, involving chemicals, specimens, and samples, while the latter lives in the cleaner, digital realm of bioinformatic data. Novak describes his role on the "wet" side as being exactly what you imagine when you think of a stereotypical scientist: "that person at a bench with pipettes and tubes and chemicals, working in a lab with beakers and devilishly laughing at these amazing reactions! Actually," he adds, "it's important not to laugh when working with the ancient DNA. You don't want to contaminate it."

A lot of Novak's work involves handling the specimens and doing the DNA extractions. For the passenger pigeon study, they've been using a large sample from the Royal Ontario Museum in Toronto. But when I saw those birds up close in real life, they looked complete to me. They weren't missing any limbs, at least from what I could tell. Novak says that's because scientists are only minimally destructive when they take what they need from them. It's just a tiny piece of the toe that's cut off—about the size of a pinhead. "In the lab I'll cut that in half, dissolve it up in enzymes, destroy the proteins, filter out the DNA, and I'll start working with the DNA from there to get it to a level where it can be sequenced." Once it is sequenced, the resulting DNA letters are sent for processing in the dry lab, where the detective work really unfolds.

The sequences are handled in the dry lab by other scientists, like André Elias Rodrigues Soares, a bioinformatician who helped Novak and Shapiro piece the extinct pigeon genome back together in the right order. They were able to sequence all of the passenger pigeon's mitochondrial DNA and assemble it with newer, improved technology since the days of working with dodo DNA. Importantly, they were also able to sequence complete nuclear genomes from different passenger pigeons, which gave them two ways to look at the species and understand its evolutionary past. But the passenger pigeon had vanished long before scientists had the ability to map its complete genome or were even interested in doing so. As a result, there are no whole, intact living cells from which to read its genetic map. So just as Hendrik Poinar's and Vincent Lynch's groups had to use elephant genomes to guide the assembly of the woolly mammoth's genome, Novak, Soares, and Shapiro have had to use the genome of a close living relative—the band-tailed pigeon—to assemble the extinct genome they're after.

Unlike the passenger pigeon, the band-tailed pigeon still soars across the Americas from western Canada south to Mexico. But no one had yet sequenced its genome, so researchers had to assemble that as well. By handing the DNA sequences that they recovered from passenger pigeon tissue samples to a start-up cofounded by Shapiro's partner, Ed Green, they were able to assemble the band-tailed pigeon genome by March 2015. Green's business, Dovetail Genomics, has developed an in vitro method for assembling DNA sequences with unprecedented speed, and the company is assisting Revive & Restore in a variety of its de-extinction projects. The next step is to compare these two related pigeon genomes to identify places in them where the birds are different from each other and where they are the same. There are long spans of the genomes in which the DNA of the two birds is identical, which Novak tells me is about 97 percent of the genome, and then there are regions where the DNA is what scientists call "highly diverged," meaning that at some point during the evolution of these two species, each bird accumulated its own mutations that, for example, make one species purple and the other blue, or that give one a brick-shaped tail and the other a dove-shaped tail. There are, however, tons of places in the genomes where the two birds differ, so it is anything but easy to determine just which differences matter in making a bird look and act more like a passenger pigeon than a band-tailed pigeon. "We do know," Novak says, "that these regions that are evolving differently between the species are creating the differences between the birds." Next, he'll have to figure out what those genetic differences between the two species do to affect the unique functions of each type of bird.

Once he understands which genetic differences matter for making the passenger pigeon different from the band-tailed pigeon, the next step will be to edit the band-tailed pigeon genome so that it includes the passenger pigeon genes he wants and then create a real, breathing, flapping bird from it—a bird that looks, and acts, like a passenger pigeon. Scientifically, it's a complicated process no one has yet come close to doing with birds, but that has not stopped Novak from planning it out. And planning is needed: because of some features of their reproductive system, which I'll get to, it is extremely difficult to clone birds.

Novak plans to use an ingenious method for reprogramming pigeon DNA in an embryo that will produce a genetically modified hatchling. First, he must isolate and harvest cultures of cells called primordial germ cells (PGCs) from a band-tailed pigeon. PGCs are predecessors of the germ cells that give rise to sperm and eggs.

To get their first experimental batch of PGCs, Revive & Restore made a deal with Crystal Bioscience, a company with expertise in culturing chicken PGCs for applications in medical research. Crystal Bioscience's role was to optimize PGC harvesting methods from pigeons. Novak says they're trying first with rock pigeon PGCs instead of band-tailed pigeon PGCs for the simple reason of supply and demand: there are a lot more rock pigeons around than there are band-tailed pigeons in captivity, and many have been domesticated. Depending on the breed, Novak tells me he's seen rock pigeons go for as little as $5 to $20 per bird, while a band-tailed pigeon can cost anywhere from $100 to $250 each. Eventually, though, they'll need more partners to fund the steps required to obtain active cultures of band-tailed pigeon PGCs, in order to get the most accurate results.

Researchers get PGCs out by cutting a little window in the fertilized egg's shell at a spot known as the *germinal crescent,* where these cells are naturally found. The PGCs are then isolated from the embryo at an early stage and grown in culture so that they can have their genomes edited. The PGC harvesting destroys the embryo in the process, annulling the potential bird, which creates yet another ethical dilemma. When Novak and I last spoke, they'd made good progress on how to best handle pigeon eggs for PGC harvesting, but there was still a lot of work to do.

One day, a retired quantitative geneticist got in touch with Revive & Restore with hopes of getting involved with the Great Comeback. He had a long career history of working with birds and was now retired, so he had not only the special knowledge but the time needed to get the experimental birds to lay eggs. Revive & Restore decided to bring him on to the project, and with his help set up a facility that houses two breeding pairs of band-tailed pigeons that Novak did research with for his master's. (One of the females, Sally, is the bird they took tissue and blood samples from to assemble the band-tailed pigeon genome.)

The anonymous retiree is doing something nifty to get these birds to reproduce: "double-clutching" the eggs they lay and getting them to breed out of season. To do that, after the eggs have been laid, he transfers them to rock pigeons, which stimulates the female band-tailed pigeon to produce another egg. He also alters their exposure to light, giving them twelve hours of an artificial source each day. This gets the band-tailed pigeons to lay eggs in January, even though they don't actually breed until April. That way the birds can provide a constant stream of embryos to Crystal Bioscience for the PGC-harvesting research. A pilot study with this method wrapped in 2015.

If and when Crystal Bioscience perfects its methods for culturing pigeon PGCs, Novak will begin his engineering work. At that point, using CRISPR, he will start to introduce synthesized passenger pigeon genes of choice into the band-tailed PGCs— genes they believe are important to making a passenger pigeon a unique kind of bird. Little by little, with each additional edit, the band-tailed pigeon genome will start to resemble the passenger pigeon genome, according to Novak's design.

"So getting one gene in, that's cool," he tells me. "But then we need to introduce another gene, and another gene, and so on and so forth until we get some cells that are carrying what can be considered (more or less) the genetic code of a passenger pigeon, and no longer the genetic code of a band-tailed pigeon." Then he must inject those edited PGCs into an embryo that comes from a female band-tailed pigeon.

To do that, he'll hijack an embryo (which looks like what you expect a normal pigeon egg would look like) at a precise moment after it has been laid: the embryo must be at just the right developmental stage to allow the injected PGCs to circulate in its blood. He'll then cut a small window opening in its shell, revealing a space he will inject the edited PGCs into using a glass capillary needle. After he has injected the edited PGCs, he'll seal up the window and incubate the embryo. As it is incubating, the edited PGCs will move through the embryo's bloodstream, ending up in the developing bird's gonads. Then, after all of Novak's hard work and the embryo's endurance, the engineered embryo will hatch into... a band-tailed pigeon.

Sorry, come again?

"Now you're going to get something called a chimera," Novak explains, "and when this bird hatches, it is not a passenger pigeon. It looks like a band-tailed pigeon, but it carries a secret in the ovaries or the testicles, where it will have sperm or eggs of these cells that we injected." The PGCs that Novak puts into the embryo will develop into the new chimera's sperm and eggs. The magic of this approach is that those sperm and eggs will contain genomes that carry passenger pigeon DNA. "So if you create a male chimera and a female chimera and let them fall in love and have some babies, their offspring will be those that we genetically created. That is, if everything we do goes really well."

Changing the game of avian reproductive science, achieving de-extinction, and having it go really well is a tall order. But Novak has time to get there and says this is decades-long work. It might take years just to develop the culture of PGCs he needs in the lab—they don't often have high survival rates outside of the bodies that first made them, and that obstacle still needs to be overcome. Another challenge is that cultured PGCs can be transferred only into birds of the same sex. A female bird can't be successfully injected with PGCs of a male bird, for example. So to eventually obtain engineered passenger pigeons, Novak needs germ cell lines from each sex that work like well-oiled machines. Getting all of that working is not something that can be rushed.

At first, it sounds like it must be easier to just do cloning, but some major roadblocks in bird biology make working with the cells in this way the most promising method for what Novak wants to achieve.

There are lots of good reasons why it is extremely difficult to clone birds. Picture a bird egg. Maybe you even ate one today. What first comes to mind? For me, it's that they're huge! Their size obviously varies by species, but compared with a mammalian egg, they're colossal. These large orbs are filled with fatty yolk and other opaque bits that make it hard to see the nucleus. In cloning, you need to know exactly where the nucleus lies—be able to see it—because you have to yank it out and transfer it into a new egg that has had its nucleus removed. Initial research has been done to make the nucleus in a chicken egg detectable to the human eye, but it remains tricky.

Another difficulty is that female birds don't have a uterus like mammalian females do. When a mammalian embryo implants itself in a uterus, it gets nice and cozy and stays in one spot. Because of this, you can develop an engineered embryo in the lab, implant it in a surrogate mother's uterus, and watch it grow. But when a bird lays an egg, that embryo is already very much developed. A bird's uterus—called an *oviduct*—is more like a conveyor belt, along which the developing egg is always on the move. If you had X-ray vision and looked at a bird from the side, you'd see that conveyor belt carrying a procession of objects that start as tiny blobs and end up as fully formed eggs with shells before they leave the bird. So it is hard to be sure when to grab the egg, take it out, and splice its DNA with the genes you choose. The point at which a bird egg is just one single cell and its nucleus is retrievable is pretty much impossible to ascertain. "Not to mention a surgical nightmare when trying to implant the egg back into the mother!" Novak tells me.

The Route to Flying Freely

DESPITE THE HURDLES, let's say it all works, and new birds are created one day that look convincing enough to call "passenger pigeons." At first there will be one newly engineered bird, then two, three, four, until finally a tiny flock is created. Where will these new hatchlings live? To that Novak responds, "Well, my vision of how this can happen is that we will give them a natural habitat as best as we can. Rather than have them grow up in an experimental box that we monitor with cameras, since our end goal is producing a viable wild population, the design we need is to build something much like a zoo enclosure—a nice forest-looking aviary that these chimeras are breeding in—where they can forage for acorns and nuts off the ground."

When the baby "passenger pigeons" are born in the aviary, they will hatch from a band-tailed pigeon egg and they will see band-tailed pigeon parents. But band-tailed pigeons and passenger pigeons are each unique birds with unique behaviors. Are band-tailed pigeons going to be able to raise a passenger pigeon? Shapiro says that "one would have to figure out a way to make that bird, when it was born, look and act like a passenger pigeon instead of like a band-tailed pigeon." "But how might someone do that?" I ask.

"That is an excellent question, and we do not know the answer. I don't know what the behavioral differences that distinguish a band-tailed pigeon from a passenger pigeon are, and this is something that would be important when trying to select which genes to actually change, or what kind of environment one would need to establish in captivity in order to raise the next generation of slightly manipulated band-tailed pigeons. There is more to making a species look and act like that particular species than just its genome sequence."

Novak has a plan to tackle the problem. The plan depends on making changes to his band-tailed pigeons' genomes based on what they are exposed to in their environment, an area of science known as *epigenetics*. "One thing I've really wanted to do since the beginning of the project," he explains, "is to take band-tailed pigeons, get them into passenger pigeon habitat, and feed them a passenger pigeon diet. The idea is that their epigenetics will change in the environment and that the birds raised out there will make better surrogate parents for the new passenger pigeons and better primordial germ cell donors for the cells that we want to engineer. They will be the cell donors, as well as the birds we make chimeras from, as well as the surrogate parents."

Novak believes that the passenger pigeon's epigenetics were flexible to environmental change, and says that the last straggling passenger pigeons were often seen with mourning doves or rock pigeons, which could have affected them in any number of ways. There's no way of knowing the extent of influence other pigeons had on them. After all, no one was studying passenger pigeons in the 1890s as we are today. But Novak can say with certainty that the first recreated passenger pigeon "might end up getting picked up by a cumbersome inquisitive monkey, such as myself, just to make sure it is healthy, so we can show the world that we've done something amazing." While the goal is to try to raise the birds as much like passenger pigeons as possible, there's no denying that the first batch will be quite special. Novak smiles at the thought, then says, "They're going to be quite the celebrity birds."

In this hypothetical world that is home to a captive population of new passenger pigeons, researchers will set up a few different flocks and choose one flock out of the group to breed a bit more rapidly by double-clutching. When the eggs that are taken away hatch, the baby birds will go to surrogate rock pigeon or band-tailed pigeon parents so that the breeding pigeons can quickly make more eggs. Some of those birds will be moved into a flock that isn't double-clutched—that produce just one baby a year, as is normal for them. Eventually, between the two types of breeding groups, there will be enough new offspring to stabilize population growth.

Novak imagines that once there are hundreds to thousands of birds, they will be ready for what is called a *soft release*. Large aviaries would enclose vast sections of natural forest in the passenger pigeon's native habitat, fenced in to protect the birds from predators, like bobcats. The wilderness is, of course, full of other wild things too, like diseases and bacteria. These are tricky to simulate in captivity, so the birds' immunity against them needs to be tested in a controlled-exposure setting. "Of course, we don't want all our birds to die," Novak says, "but we need to put them into this wild setting to know if these birds have a chance to survive at all." If that works, then over time, other animals that the original passenger pigeons once interacted with will be let into the enclosure—small creatures like chipmunks, mainly, and blue jays. At that point, Novak will sit back and observe. And after a long period of monitoring, if the birds look like they are adapting healthily to the changes around them, he'll take them on their first flight. The plan is to fly them from one aviary to another until they are able to mimic the passenger pigeon's nomadic lifestyle. But for this, Novak needs the help of a gaggle of flight specialists he calls the Pigeoners.

One day, a man from Ohio called Shapiro's lab and left a message expressing interest in hearing more about the passenger pigeon project. Novak returned his call expecting to hear a fearful naysayer on the other end of the line. It's happened a few times before. "Once," he tells me, "we got this letter that said, 'Stop now and turn back before your monster pigeon destroys the world!' And I was just like, 'It's a pigeon, you know?' " But to his surprise, Novak heard the kind of excitement in the man's voice he recognizes in himself.

It turns out that the caller, who had previously won an award for racing pigeons, had seen a video online of one of Novak's talks. He was struck by Novak's description of his plan to fly the birds between different facilities with surrogate flocks and knew he could help make that real. Novak told him that it might be ten years before they do something like that, but the man's interest didn't fade. Ever since, he's been helping Revive & Restore assemble their network of Pigeoners like him—people in Ontario, New Mexico, Minnesota, Kentucky, Massachusetts, and his own state of Ohio. They want to eventually build pigeon facilities in each of those places and run regular flights between them. Only after scrupulously watching those flight tests and ensuring that they are successful can Novak imagine starting to introduce new passenger pigeons into the wild. "If it looks good, you take down the aviary walls and say, 'Okay, this first generation of very promising passenger pigeons that have been exposed to the wild, they seem to look like they know what they're doing, let's let 'em go and let them be free.'"

It sounds simple when he puts it so nonchalantly, but fears lurk in the shadows of his mind. "Oh, my biggest nightmare, I think it's going to be very, very tough at any stage of the project if something just seems impossible. If the germ cells just don't culture, if the genetic engineering just doesn't happen. But a bigger nightmare would be if, you know, in some horrible way, they end up being sterile. Or if all of that goes well, and then the first promising birds die, that's going to be a tough moment." There's not much he can do to prepare for that sort of potential devastation except remain hopeful in the face of uncertainty. Early on in my research, it became obvious to me that you have to be a glass-half-full type of person to do this work.

But listening to Novak go on in his self-assured way, I had to pinch myself. Passenger pigeons? Huge flocks of them? More pigeons than we already have? Is he for real?

One of the biggest issues with reintroducing the passenger pigeon—or some facsimile of it—into North America is that the species could potentially roost again in unfathomably large flocks and become a frightening pest. Imagine billions of pigeons flying above our cities and towns, the thunderous clap of their copious flapping wings impossible to ignore. And although it is thrilling—and even scary—to imagine their abundant swarms, there is no scientific evidence that says we should expect recreated pigeons to return in the same numbers. There isn't any precise way to test if they will, since ecosystem dynamics will play a role.

However, it has been speculated that something called the Allee effect might be crucial to the passenger pigeon's fitness and therefore its abundance. The Allee effect is found whenever a correlation exists between a species' population density and the ability of an individual in that population to survive. It has been hypothesized that passenger pigeons needed huge numbers of their fellow birds around them to breed, build nests, care for their young, and defend themselves. Population studies show that there were still vast numbers of the birds up until the late nineteenth century but that they rapidly crashed to extinction a short few decades later. Some have posited that once they fell below the critical Allee-effect number specific to their population, it was free fall until they were all gone. If that is true, then no heroic efforts made now to bring back a handful of passenger pigeons would be worth it. If these birds truly couldn't survive without their enormous flocks, then making a few in a lab now would simply be futile. How long would those first few be able to live before the Allee effect took its toll? But Novak happily told me on a couple of occasions that this doesn't seem to be the case. Recent insights made from genetic studies of the bird suggest that the Allee effect may not apply here at all.

Then, in December of 2016, he told me something else. New studies of the bird suggest that its genome is actually *built* for an Allee effect. "Now, the complication in considering this topic for passenger pigeon de-extinction," he explained, "is the idea that we are somehow recreating the passenger pigeon as it was—which is not true. We are reshaping a band-tailed pigeon." Since the band-tailed pigeon appears to not be built for an Allee effect, Novak says, his unextinct passenger pigeons—which will be made mostly from the band-tailed pigeon's genetics—should not suffer from this effect for many generations. If they ever do, he postulates, it will only be after the birds reach incredibly high levels of bioabundance, which will take a lot of time. "In a thousand years the recreated passenger pigeons could look just like the extinct genome, and then an inbreeding Allee effect would pose a challenge to them if they were to ever go through a population crash again—but this is the sort of thing that careful management and captive breeding can avoid." No matter what surprises the science throws at him, Novak always finds a way to remain sure-footed and hopeful about bringing the bird back.

Ever since he started working for Revive & Restore, Novak has been pecking away at a master's thesis that looks at the ecology of passenger pigeon habitat in relation to its evolutionary past, a thesis he successfully defended in March 2016. His graduate work shows that the passenger pigeon habitat shifted much more often than the population did. "Not only was there a lot of them," he told me, "but they were able to maintain abundance no matter how the environment changed." He added, "The passenger pigeon can be summed up with two concepts: (1) ecologically resilient (an environmental super-species), not prone to extinction by natural earthly processes, and (2) genetically tied to living in abundance, and therefore vulnerable to catastrophic events (such as a hungry American population armed with nets and guns)."

If those conclusions are correct, they will have major implications for the bird's de-extinction. Some critics have postulated that the northeastern American forests where the pigeons used to live have changed too much since the species disappeared to be able to support flocks of billions today. The Carolina parakeet, the Bachman's warbler, and the ivory-billed woodpecker were all dwellers of the same trees that vanished around the same time that the forests transitioned. By about 1870, the extensive deforestation of the northeastern United States by European settlers had reached a point where nearly 50 percent of the forest had been clearcut. By connecting the dots now between the disappearance of various species and the massive deforestation efforts of the last century, a story of extinction that doesn't implicate hunting alone becomes visible. So the idea that those same forests would somehow be suitable today as habitat for enormous numbers of new birds is, at first glance, not so convincing.

Although many people say that the passenger pigeon's habitat—as it once supported them—is now gone and therefore new unextinct passenger pigeons would have nowhere to go, Novak says they're wrong, a claim that his graduate research backs up. In an interview with Mark Peck and his colleague Dave Ireland from the Royal Ontario Museum, he said, "The reality of the passenger pigeon is that it has been around for millions of years... The species has been what it is for a very long time. That's through ice ages and drastic forest changes, that's through forests dominated by pine trees to complete forests dominated by oak. Regimes change, this bird lives through them." Therefore, the idea that the forests are now too different to support them violates the crux of what the passenger pigeon was so well adapted to do: withstand environmental change.

"The fact of the matter is," he continued, "that at present, in the eastern United States there are more acres of forest now than there were in 1890. This is a rare opportunity for an environment in which we have more habitat rather than less... And it is a habitat that can benefit from having its old levels of biodiversity back. One crucial element of that biodiversity was an arboreal wood pigeon—the passenger pigeon." Because the passenger pigeon ate nuts, seeds, and berries, then flew 500 miles or more away to deposit those nutrients elsewhere, they regenerated vast landscapes in a dynamic way. That's why Novak says their flocks acted like a "super-organism" and profoundly affected the ecosystem similarly to how megafauna of the Pleistocene did when they were alive.

He hopes for the recreated passenger pigeons to compete with "mice, deer, squirrels, blue jays—allowing biodiversity to kind of balance out and become enriched." No one knows yet how many birds will be needed to do that, or for them to act like a "super-organism," flocking densely enough to play out the nomadic, disruptive, and regenerative role for forests that passenger pigeons once did. Novak said in the interview with Peck, "People get scared of the idea of large flocks, but the idea of a large flock for the future is not necessarily what it was in the past... Maybe a billion? Maybe a little more? What does that mean for people? Well, they're not necessarily going to see all those pigeons in one spot all the time; and they are not necessarily going to see them often." It depends on where you live, and where the birds decide to roost.

When they were alive, the nesting flocks of billions of passenger pigeons broke tree branches and filled eastern forests with their droppings, disturbing the ecosystem in ways it wouldn't have been if passenger pigeons had never been around. Part of Novak's argument is that forest disturbance as a byproduct of how abundant the species was affected the forest in beneficial ways that have been lost since the bird went extinct. Therefore, flocks that are large enough to cause forest disturbance—even flocks not as large as they once were—are exactly why he says we need new passenger pigeons in the world now. Revive & Restore's website describes the rationale: "Passenger pigeons and fire were the major sources of beneficial and continual forest disturbances for tens of thousands of years. Since the extinction of the passenger pigeon humans have suppressed fires, leaving no consistent source of forest disturbance. While the eastern United States has experienced vast reforestation over the past seventy-five years, regeneration has virtually stopped, leaving new and old forest stands stagnant and native species on the decline. By reviving the ecological role of passenger pigeons we can restore and perpetuate forest regeneration cycles naturally."

Novak once broke it down very cogently for me: "I mean, why bring back a passenger pigeon? It plays a particular role in its environment, a role we feel is valuable, and this is forest disturbance. These giant flocks disturbed canopies and consumed a lot of nuts. They did a lot of things that would be equivalent to a hailstorm or a fire." When hailstorms or fire disturb forests, the places where they remove tree bark and canopy, or scorch roots and trunks, can spur those sites to produce sprouts and even regenerate seeds. Speaking of hailstorms and fire, Novak added, "That sounds terrifying, I know, but there are a whole bunch of species alive today—animals like the New England cottontail or the American woodcock—that thrive on first-generation forests that pop up after these disturbances. There are management plans trying to save these species that involve having to recreate forest disturbances. And if I had to choose between a forest fire, a hailstorm, or some pigeons eating a whole bunch of nuts, I'm going to choose the pigeons."

But many people I've spoken with would choose otherwise. After all, more flocks of pigeons would mean more bird droppings on the hoods of their cars.

For a man who thinks about passenger pigeons so ardently, Novak doesn't think about the most famous one—Martha, the last passenger pigeon—all that much at all. At 1 p.m. on September 14, 1914, Martha died at the Cincinnati Zoo. She didn't simply croak but withered away inside of a cage whose bars she had been watched through for most of her life. Novak sees Martha as an icon of North American extinction, but he is not encouraged by how she lived or what she symbolizes for her species. Martha spent her entire life in captive flocks and died without ever having successfully bred. She came from a stock that may have had only a single breeding pair of pigeons to begin with, meaning she might have been not only bred in captivity but also *inbred* in captivity. Today, she sits in the pits of the Smithsonian Institution's National Museum of Natural History in Washington, DC. She is placed on a branch that is screwed to a second branch that has another passenger pigeon perched on it. I've not seen her myself in person, but Novak tells me she isn't the most attractive mount: her tail feathers are in rough shape; she looks a bit disheveled. And when he stares at her, he sees the death of a species rather than the spirit of its impressive return. For him, what has more impact than seeing Martha is seeing the passenger pigeons that nobody really knows or cares about. "When you see a drawer full of these stiff, almost distorted museum skins," he tells me, "that's when you get the full powerful drop-kick in your gut that what was once an entire, breathing beautiful species is a drawer now. It's a dust-collecting drawer of lifelessness."

Novak's personal records show that he has already viewed up to 178 passenger pigeons in collection drawers or display cases at 26 institutions in three countries. His goal is to see preserved passenger pigeons in at least two more countries by 2020.

Novak wrote an almost spiritual blog post on Revive & Restore's website in 2014, for the one-hundredth anniversary of the passenger pigeon's extinction. It read, "A new passenger pigeon may cause an internal debate in the on-looker, but when the piercing red iris meets the on-looker's gaze, in a moment of deep connection, it will remove all doubt, as if to say 'I am the passenger pigeon reborn.' The spirit of this animal will be as vibrant as Martha. Unlike Martha, this new bird's story doesn't have to end in a cage. With a generation behind it, these new birds can grace the forests again and enrich the dancing cycles of the trees; they will enrich our souls if we open our hearts to them... We ultimately write the story... Her story is far from over." Humans more broadly may ultimately write the story, but Novak is the primary author. And after years of following his progress, I still have no clear idea of where his well-intentioned plotline will lead us. All I know is that I'm quite a bit more apprehensive about it than he is.

ONE OF THE last times I spoke to Novak for my research, he answered the video call wearing a gray rubber pigeon mask that covered his entire head, sending me into a fit of laughter. He then started flapping his arms like wings, which I found even more hilarious, and I could only manage to nod in response. This is Novak in his element. He is a fun-loving and energetic young scientist whose unabashed enthusiasm for the pigeon that holds his affection is absolutely infectious, if maybe also a little odd. He was getting ready for a trip to Geelong, Australia, where he was going to receive training in gene editing in bird embryos. Then, in December 2016, he received the good news that he had been awarded scholarships that will allow him to move to Australia in 2017 to do a PhD at Monash University, Melbourne. Down under, he'll be doing genome editing in rock pigeons to begin modeling the future of passenger pigeon de-extinction and contributing to other bird genomics projects.

Over the time that Novak and I have been talking about his work, it has become increasingly evident that the nascent science behind the Great Passenger Pigeon Comeback is bursting through its shell. Indeed, Revive & Restore has now been able to publicly declare when it expects to reach its goal. The plan is to hatch the first generations of passenger pigeons by 2022 and try to reintroduce them into the wild about ten years later. On that long schedule, it's pretty clear that Novak will not follow his and my millennial generation's trend of having multiple different careers in one's lifetime.

He tells me that by the time this book is published, Revive & Restore will have even more avian research to share. In 2017 they're setting up a Revive & Restore Genetic Rescue Service independent from the Long Now Foundation that currently houses it, where they'll establish genomics and biobanking projects for over a dozen climate-threatened North American birds, as well as investigate the genomes of extinct peregrine falcon subspecies. "Birds certainly seem to be driving the future of genomic conservation, despite their challenges!" he wrote in an email updating me about their fast-growing work. But even before Revive & Restore fledges from the Long Now nest, they have begun incubating hope that another culturally significant extinct bird could come back to life.

De-extinction Flies Further Afield

AS THE STORY goes, the bird eaten by the pilgrims at the first Thanksgiving dinner—the heath hen—lived in the shrub and grasslands of New England and its surrounding areas. Like the passenger pigeon, the heath hen was easy to catch and tasted good, and it too was driven to extinction by market hunting. After it disappeared, the ecological productivity of its original grassland habitat became depleted. Revive & Restore would like to see that habitat bounce back. To that end, members of the organization are working with the community in Martha's Vineyard—where the last individual heath hen, Booming Ben, died in the spring of 1932—to raise the funds and social acceptance needed to bring a recreated heath hen back to the area.

After garnering financial support from a handful of residents of Martha's Vineyard, Revive & Restore partnered again with Dovetail Genomics, this time to assemble a high-quality template genome of the heath hen's closest living relative, the greater prairie chicken. Using fragments of heath hen DNA from preserved birds in museums in Chicago, Philadelphia, and Toronto, researchers mapped the extinct bird's genome onto the template genome. They then compared the heath hen's genes to genetic sequences from its other living relatives, like the Attwater's prairie chicken, the sharp-tailed grouse, and the lesser prairie chicken, to see if the heath hen is genetically distinct enough from its living relatives to warrant its revival. If there are too few genetic differences between the living species and the extinct bird, its existing relatives could play out the ecological roles researchers want to see restored, and de-extinction of the heath hen would be pointless.

But the results showed that the heath hen's genome *is* distinct enough from its living relatives' genomes to assume that it once had a unique environmental role to play. As a result of this finding, Revive & Restore has pushed ahead with the revival plan. One of the most earnest local supporters of the project is Teddy Palmer, who, when he was eight years old, donated $70 in one-dollar bills from his lemonade-stand profits in exchange for a promise that he can hold the first unextinct heath hen when it is born.

The passenger pigeon and heath hen revival projects are moving forward simultaneously so that any advances made in one can benefit the other. In both cases, the factor that most limits progress will be the ability to do genome editing in avian primordial germ cells. But the heath hen is not all that distantly related to the chicken, which has been extensively studied, and that gives it an edge over the passenger pigeon. If Revive & Restore's method for de-extinction via primordial germ cell modification succeeds, Phelan hopes it can be adapted and applied to a vast menagerie of lost and threatened birds. That idea has not been lost on others who want de-extinction researchers to think about what other bird candidates might benefit from the progress being made on the passenger pigeon and heath hen projects.

In 2015 Matt Ridley, the British science writer and former bird biologist, organized the first de-extinction conference on the return of the great auk under the auspices of Revive & Restore and the International Centre for Life, a "science village" in Newcastle upon Tyne, England. The great auk, he believes, is a particularly good candidate for de-extinction because it is the only bird species that bred in Europe to have gone globally extinct in the last two hundred years. Once numbering in the millions, the great auk was a large, flightless black and white northern bird that looked a bit like a penguin but evolved separately, with a ribbed, hooked beak as big as its head and short stubby wings that prevented it from achieving lift-off. "If we are to achieve de-extinction, the first candidate from Europe has to be the great auk," Ridley tells me. It lived on remote islands and coastlines all over the world, including Canada, the United States, Norway, Greenland, Iceland, the Faroe Islands, Ireland, Great Britain, France, and Spain. Great auks stayed close to the sea along sloped landscapes with rocky shores that allowed them to easily reach the water. But because the species was flightless, this meant that other creatures could easily reach them too.

The Neanderthals found the bird tasty, as did we. But even more to the bird's detriment, its downy feathers made for great pillow stuffing. As a record from 1775 describes in gruesome detail, "If you come for their Feathers you do not give yourself the trouble of killing them, but lay hold of one and pluck the best of the Feathers. You then turn the poor Penguin adrift, with his skin half naked and torn off, to perish at his leisure. This is not a very humane method but it is the common practice. While you abide on this island you are in the constant practice of horrid cruelties for you not only skin them Alive, but you burn them Alive also to cook their Bodies with. You take a kettle with you into which you put a Penguin or two, you kindle a fire under it, and this fire is absolutely made of the unfortunate Penguins themselves. Their bodies being oily soon produce a Flame; there is no wood on the island." And by the mid-1800s, there were no great auks either.

The great auk's genetic sequences describe a close relation between it and a living bird: the razorbill. The razorbill, presumably, can therefore aid in the de-extinction of the great auk. As Stewart Brand wrote in a summary of Revive & Restore's activities for 2015, "The razorbill has similar range and feeding habits and an almost identical appearance, except that it is one-eighth the size and can fly—two traits that may be easy to adjust." That cavalier attitude about fixing such prominent features made me pause, and when I ask Ridley what he considers to be the most legitimate set of concerns associated with bringing back the great auk, he mentions these discrepancies: "The possibility that we may not be able to produce a creature that is precisely like a great auk, because of some genetic, developmental, or behavioral deficit that cannot be foreseen. If, for example, a bird is hatched that is not quite the right size, color, or shape but is very nearly right and is quite healthy, then can we call it a great auk or release it into the wild?" No one knows yet, but in time Ridley may try to find out.

The great auk's genome has been sequenced by Tom Gilbert at the University of Copenhagen, and Ridley has prepared a good amount of material in order to discuss the ethical, ecological, ethological, and economic challenges of getting the bird back. But so far, the great auk's de-extinction is just a paper project, not yet an experimental reality. "I have no immediate plan to take this further," he told me at the tail end of 2016, "but I hope to get to the point where it becomes possible to start a project some time in the next five or ten years"—after some other bird species show the way.

De-extinction is as much about pushing an entire science forward as it is about achieving any one species-specific goal. As the science advances in one experiment, it will be able to be applied in others. Novak gave an example of how this works in a reflection he wrote about his experiences at a conference where he presented his research: "Two years ago at the 2014 Avian Model Systems 8 conference, the only mention of CRISPR/Cas9 was in reference to its future use in passenger pigeon de-extinction—despite CRISPR/Cas9's rapidly spreading use in other model organisms at the time. At this year's meeting, every avian transgenic lab in attendance reported using CRISPR/Cas9."

Novak's project and the technologies that underpin it invigorate and inspire others to imagine how they might use similar methods for their own interests. Like evolution itself, the variables of this science will continuously mutate and produce outcomes in the future that are not yet possible to grasp. It's the ultimate case of letting the cat out of the bag—the cells out of the lab—the bird out of the hatchery. And it's going to change a lot of what's possible along with it.

Something I find striking about all three of these candidate birds—the passenger pigeon, heath hen, and great auk—is that we used to eat them all. And that raises an important question: If they were to come back now, would we not want to eat them again? At first, the idea that one might actually eat a resurrected animal sounds barbaric. After all, why would we go to all of the trouble of recreating a species if we just plan to kill it again? Seriously, we have so many other things to snack on. Then again, people thought that passenger pigeons were nutritious, tasty, and cheap, which partly explains why they're no longer here. We can point to other cases where animals have been on the brink of extinction, then saved, but not without selling some of their meat for human consumption. For example, the North American bison was at the edge of extinction at the end of the nineteenth century before it was bred back to healthy numbers in conservation programs. Most of those bison live in national parks around the U.S. today, though many are also found in North American grocery store freezer aisles as the contents of barbecue-ready burger boxes.

Select populations have been created for the sole purpose of harvesting their meat, and the bisons' story demonstrates how eating revived animals does not necessarily preclude the majority of them from living in the wild. In this sense, it becomes conceivable that flocks of revived passenger pigeons could possibly benefit several ecosystems at once. They could restore the northeastern forests, teach us about de-extinction in captive zoo populations, and nourish us. Black-and-white thinking does not reflect the realities of what de-extinction is capable of, and there may be a spectrum of outcomes for each species that scientists try to recreate. But I worry that if unextinct meat ever becomes coveted and expensive, it could be exploited as a status symbol for self-absorbed humans rather than as a source of reasonable nutrition.

Take the case of the giant salamander, a critically endangered species that can grow up to six feet long. It is a delicacy in China, where a scandal broke out when the Shenzhen police chief was caught eating one at an upscale restaurant. The endangered salamander in question had been provided by one of the other dinner guests, who allegedly bred it illegally in captivity, presumably for this purpose. Twenty-eight people at the dinner became violent when journalists showed up and tried to document their meal. "I don't know about you," Tom Gilbert says, "but I don't think that most people who would eat a giant critically endangered salamander do so because of how it tastes. People eat a giant critically endangered salamander because they can spend $15,000 on drinks to go with it." Exotic foods yield significant status in some circles and can put endangered species in even more peril than they're already in. And what could be more exotic than meat marketed as having been "brought back to life from the dead"?

CHAPTER SIX

HOW MIGHT WE REGULATE THIS NEW WILDERNESS?

De-extinction and the Law

EVERYTHING THAT CAN be written about de-extinction and the law, at the time that I'm writing this, is total speculation. There is no working example of such a law yet, so no one can say for sure how the law will apply. When Celia, the last bucardo, was briefly "brought back to life" in the form of an individual clone, no legal precedents were set, most likely because the bucardo only lived for a few minutes after it was born. The lack of success may have prevented the researchers from having to think about what might happen—legally, that is—if a herd of recreated bucardos were ever released into the wild. But since American and Canadian law tends to focus heavily on precedent, lawyers can make educated guesses about what is plausible for de-extinction based on what's already occurred. Norman Carlin, an environmental lawyer with a PhD in biology, reflects that approach in his research about the regulation of de-extinction and is therefore a great asset to any legal discussion about species revivalism.

As you know by now, it is unlikely in every imaginable case to recreate a completely genetically identical copy of an extinct species—identical in both its nuclear and its mitochondrial DNA. That will change only when we are able to synthesize full genomes from scratch with 100 percent accuracy and when the process becomes cheaper. But even then there will be no womb from the extinct species for the unextinct specimen to develop in, so differences are unavoidable.

This is why Carlin uses the term *facsimiles* to refer to unextinct animals. As he says, "Because any realistic prospect of recreating extinct species involves a fairly intense degree of genetic manipulation, you are dealing with something that is a product of human intervention and therefore it should be, probably, patentable under existing laws and precedents." Innovations with a clear inventive step—meaning the degree to which they are sufficiently inventive or, in other words, nonobvious— are generally patentable. A patent owner has the right to exclude other people from making, selling, or importing their invention. It's a negative monopoly that allows people to make money by preventing other people from doing something unless they pay licensing fees to the patent holder. Because, as a result of the techniques that generate them, unextinct animals are always in some way different from the original extinct species, patenting should be possible, Carlin says, and money can be made.

However, if an unextinct species somehow ends up exactly like the extinct species was before humans intervened with its genes, the law indicates that it would likely *not* be patentable: you can't patent a product of nature. For example, if Michael Archer and his team ever successfully clone the gastric-brooding frog using intact frozen cells, the clone might not be patentable—if, that is, the law considers it to be genetically identical to the extinct frog on the basis of having the same nuclear DNA. It remains to be seen whether the leftover mitochondrial DNA from the host egg cell (which would come from the great barred frog) would be considered significantly different enough from that of the extinct frog for the clone to be patentable. But there are few other similar cases: gene-edited passenger pigeons and woolly mammoths are not likely to ever have completely identical genomes to those of the extinct animals, and therefore should be patentable.

Carlin speaks from his expertise in American law when he says that all GMOs (genetically modified organisms) are patentable, but this does not apply in all nations. In the United States, it is possible to patent a *higher organism*—an antiquated term for a creature that has a certain amount of biological complexity. A classic case is the experimental animal known as the *onco-mouse,* a genetically modified mouse made in the lab specifically to be susceptible to cancer for use in research. The onco-mouse has been patented in the United States and in many other countries, but the Supreme Court of Canada has ruled that no "higher life forms" can be patented under its law. As a result, if one of George Church's mammoth–elephant hybrids ever moves up to the Canadian Arctic, whatever patent Church might hold in the United States would not apply there. But in legislative jurisdictions where patents on "higher organisms" do apply, Carlin says, there's a lot that an unextinct species patent holder could stand to gain. As de-extinction advances in the name of ecosystem restoration, might it also create an unintended market for pet companions of a sci-fi sort? Could someone holding the patent for a resurrected species like the passenger pigeon personally benefit?

When I pose this question to Carlin, he reminds me that the goal of those working on the passenger pigeon is to reintroduce it in the wild, not to market it to pigeon fanciers. But, he concedes, "you could well imagine that that approach might appear from entrepreneurs of a more financially oriented sort. The eventual experimental success is not going to be inexpensive. It might be more likely that the prospect of selling de-extincted species for profit would be a motive for investing in the technology necessary to create them." There are loads of fantastic-looking extinct species that people might go crazy for if they could get them back now as a companion. "I always cite the Carolina parakeet," Carlin says. "No one is working on resurrecting that as far as I know, but they were gorgeous birds, spectacular. You can see them in Audubon's paintings before they were wiped out in the nineteenth century. I can well imagine that those willing to pay tens of thousands of dollars for a macaw these days would pay even more for a Carolina parakeet. It would be the ultimate pet for a parrot connoisseur." Any laws that are made now relating to unextinct creatures could create serious implications for animals in the future because legislation around GMOs sometimes ripples forward through time. For example, in Europe, where the law requires GMOs to be licensed, you would need permission to create the first generation of genetically modified unextinct animals. The review process for GMO release is exhaustive, requiring that you complete hundreds of pages of detailed scientific analyses and environmental assessments about what would happen if something went wrong with the release. There's always a very real chance that an application won't be approved. Yet if you do get the license, not just the first generation of creatures you make, but *every* animal that they reproduce afterward will count as a GMO. So hypothetically, even a hundred years after getting a herd of recreated mammoths into the wild and breeding on their own, without any human management, their descendants will count as GMOs in the eyes of European law.

Meanwhile, in the U.S., if public lands or funds are part of the intended release, an environmental impact statement must be filed with a government agency. These statements are subject to public comment so that people can voice their concerns. For example, some might worry that revived passenger pigeons could become a superabundant pest flying over their neighborhoods. Future restrictions on GMOs could always arise as well, ones that we can't yet begin to predict.

Do We Have a Legal Obligation to Bring Extinct Species Back?

ANDREW TORRANCE, A professor at the University of Kansas School of Law, was asked if he would be willing to lend his legal expertise to Matt Ridley's research project in order to sort out the legal ramifications involved with the prospect of "de-extincting" the great auk. Torrance agreed to help, and started by sitting down at his desk and staring at a blank sheet of paper. Over time, on that sheet, he tried to sketch out all of the legal and ethical issues he could think of that are wrapped up in trying to revive a species. What he ended up with was a two-part framework that could answer some of the pressing legal questions.

First, Torrance compiled a review of the biotechnology regulations and laws, some of which are statutory, some of which are administrative, and some of which are informal that people just tend to adhere to in labs when they carry out particular kinds of experiments. He also listed all environmental regulations, laws, and statutes—most prominently the rules that concern invasive species and how to handle them, since a common concern about reintroducing unextinct species is that they might become invasive species in the ecosystems they are released into.

What he found out surprised and enthralled him. He had already been aware that there is a big body of existing North

American and European law dealing with *invasive* species that live elsewhere and are being brought into a new environment. But when he looked further into those rules, he discovered that there is also a large body of law dealing with the possibility of reintroducing a particular species that has gone *locally extinct* but that still exists elsewhere. To Torrance, this sounded a lot like de-extinction.

"Now," he says, "I think the implicit idea in those rules is the assumption that the species exists somewhere else and that it is just locally extinct, but in a lot of those rules they don't explicitly state that. What they say is, if there was a species there before and if it is not there now because it has been driven extinct, there is actually a positive value in bringing that species back in a way that the population could be sustained." Implicit in those rules is the contested ecological idea—one that de-extinction in the name of ecological restoration depends on—that if a species goes missing from an ecosystem, that ecosystem somehow becomes impoverished, and that therefore restoring that species to its old habitat will return the ecosystem to a more beneficial state.

In his research, Torrance unearthed a gamut of rules expressing this idea, ranging from international treaties all the way down to agency regulations. He finds these fascinating because they force one to ask the question, what does it mean that the creature is locally extinct? Based on these decrees, Torrance wonders, "Could you argue, for example, that bringing back an organism that was in existence about 100 years ago, like the passenger pigeon, is a legitimate application of the rule? Because it is extinct but was part of the ecosystem very recently, is there an affirmative obligation to bring it back to the ecosystem?... These are gripping questions that are only being raised because technology has pushed existing laws beyond where the existing laws were meant to go."

It's a puzzling thought, really: *Should* laws that were written at a time before de-extinction was possible be interpreted and applied to ecological reintroductions of unextinct species today and in the future? The drafters of these laws could not have contemplated the global extinction of species that might later be reintroduced into certain ecosystems after a process of technological revival. So what might the courts do now with those already established rules that say returning locally extinct species to their native areas is a good thing, and ought to be done if it can? "Normatively, I'm not saying that this is a good or a bad thing," Torrance tells me, "but the philosophy behind it can actually be quite compelling if you think that a globally extinct species contributed something to an ecosystem that is required. If so, then I would argue that the government should help them and protect them and, you know, enhance efforts aimed at bringing a lost species back that could complete the ecosystem." Torrance says that the argument becomes more attenuated with a species like the woolly mammoth, which has been gone for thousands of years as opposed to just over one hundred years like the passenger pigeon, though it remains to be seen by how much.

But what if people decide that they don't want these pre-existing guidelines to apply to de-extinction and argue that applying them would be a silly extension of the law? Can brand-new laws be written for reintroductions of unextinct species, ones that do not adhere to the same rules? "Absolutely," Torrance tells me. "The courts could simply say, 'Look, this law only extends to species that are still alive somewhere and don't have to be genetically revived.' In the U.S., Congress could say, 'This stuff is freaky, crazy stuff and really makes the public feel uneasy so we are going to make a new law that says none of our pre-existing pro-release laws apply to de-extinct creatures—they only apply to creatures that are unmodified and still exist somewhere in the world.' "

Either of those things *could* happen, but Torrance emphasizes that it would require an awful lot of activism to make it so. It would likely take a disaster to galvanize enough energy because lawmaking bodies like the U.S. Congress face so many more pressing issues. The trade and crime agendas, for example, would likely come first. "If some people call for statutory change," Torrance says, "Congress might say, 'Yeah, yeah, yeah, great idea, put it there on the list at position 501 and we'll get to it in ten years." And in the meantime, while we're waiting, advances in de-extinction will simply march forward. Torrance has the feeling that, as a result, we will end up largely modeling the regulation of de-extinction on the laws we already have.

Protecting Unextinct Animals from Us, and Us from Them

WHEN SOMEONE FINALLY succeeds in recreating an extinct species, the population will be extremely vulnerable to re-extinction if any of the first few die. And if they're returned to the wild, they're going to need protection from people who like to hunt animals or may be keen to see them go extinct again. But for an unextinct population to be considered an endangered species, someone must list it as such with a government body. In the United States, that would take a petition to the U.S. Fish and Wildlife Service or the National Marine Fisheries Service. A service committee would then have to debate the legitimacy of the proposal before it makes a decision about giving the species endangered status. And since an unextinct species is not an exact replica of the extinct species but one with special ontological status, it is not yet clear how it would be listed under the Endangered Species Act. Depending on how the species was created, it might be best to list it as a distinct population segment of the living species from which the facsimile was genetically engineered, rather than as a distinct population segment of the extinct species itself. Norman Carlin suggests, for example, that the revived passenger pigeon would be more suitably listed as a distinct population of the band-tailed pigeon than as a related population of the passenger pigeon because it would mainly contain band-tailed pigeon DNA, with only bits of passenger pigeon DNA spliced into it.

However, there's no guarantee that any genetically modified creature will ever receive endangered species status in the first place. If you file for a recreated woolly mammoth to be listed as endangered in the U.S., you will have to show that your newly unextinct species is indeed in danger of going extinct and that it qualifies as one of the populations or subpopulations that the Endangered Species Act recognizes. Even so, decision makers could simply refuse the request. Throughout the history of the United States, no one has ever seen a woolly mammoth walking around. It could be denied status as a native species on those grounds alone and precluded from being listed. More disconcertingly, regulators might even go one step further to call it an invasive species. As Torrance says, "Any de-extincted organism could be considered an invasive species, and in some ways, it literally is one. It is something that wasn't here in historical times and we've released it just like we've released the zebra mussel and purple loosestrife." Although zebra mussels and purple loosestrife were unintentionally released, it is possible that the intentional release of unextinct organisms could also cause invasive spreading.

It can be hard enough to get a native, nonmodified species listed as endangered, Torrance tells me, so he is skeptical that it would be easy to get something as novel as a recreated mammoth on the endangered species list. Although the Endangered Species Act does have a program for protecting wildlife bred in captivity, the snag with de-extinction is that there are no unextinct species already living in the wild. Therefore, Torrance says, it might be too much of a legal stretch to grant unextinct species protective benefits under the guise of being "wildlife" bred in captivity.

Discrepancies will likely arise across international conventions, and it is unclear how unextinct species will be handled in each legislative case. For example, under the Cartagena Protocol on Biosafety, which basically restricts how GMOs can cross borders, some decisions will have to be made about whether or not unextinct animals are GMOs or natural varieties. If the unextinct creatures are arguably GMOs of some kind, and if a bunch of them are released in one country legally but are likely to cross a border into a neighboring nation, releasing them in the first place may end up violating the Cartagena Protocol unless the country releasing the animals has the permission of the neighboring country for the animals to cross its borders. This could create some serious political tensions. On the one hand, depending on the environmental laws of the invaded country, that country may have to protect an invading species because it is considered an *endangered* species. On the other hand, under the Cartagena Protocol, if these animals enter without a license, it might be suggested that they should be killed because they are members of an *invasive* species. Torrance brings the conflict to life: "So you can imagine an enforcement agent trying to figure this out, saying, 'Wait, hold on, do I have to protect it or do I have to exterminate it because there are conflicting laws here?' "

Even within a nation, there may be inconsistencies in management, as has already occurred with the banteng—the world's second endangered species to be cloned. It is not considered part of the native banteng population in the eyes of the U.S. Fish and Wildlife Service, which instead considers it a hybrid. However, the Association of Zoos and Aquariums treats the (now deceased) cloned banteng as part of the native species population and keeps records of it in the same studbook as the non-cloned bantengs.

There's also the issue that listing a species as endangered creates a set of serious constraints in how people can interact with it. It is a lot easier for people to deal with animals on their own terms when they're not listed as endangered, so much so that it can deter people from listing a species in the first place. In the U.S., if an endangered species lives on your land, the authorities can walk in and tell you not to do something. It is prohibited to hunt, harass, or kill endangered animals or to seriously interfere with their habitat. After all, that's the whole point of listing a species for protection. But this leads to a conundrum of whether or not to list, and conservation is no stranger to perverse incentives. Once a species gets listed, it becomes infinitely valuable in the eyes of the Endangered Species Act, which may open a Pandora's box of unintended consequences. As Josh Donlan, a conservation scientist and director of the organization Advanced Conservation Strategies says, listing species may actually give someone the nudge they need to remove the species from an area altogether. If the government comes in and demands that you stop logging or fishing because you're killing an endangered species in the process, then your hands are tied. But clever people can easily figure out how to loosen the knot. "There's a saying for that in the western U.S.," Donlan tells me: "Shoot, shovel, and shut up." De-extinction may become driven by corrupt motives in some cases, but conservation is already full of them.

Again in the U.S., if the species is not endangered and if you're a private citizen, you can genetically modify an organism and release it on your own land—as long as you don't live in a particular jurisdiction that has legal ordinances against GMOs. For instance, in Monterey, California, county law says that you can't release any GMOs, so it would be illegal to do any rewilding that uses unextinct animals there that had their genes tweaked in a lab. But in most of the country, private citizens can do as they please on their own property. It only gets tricky when the animals pose a threat, such as the possibility that they might spread disease to other animals—a neighboring farm's cows, for instance. In cases like that, laws that restrict where animals can wander might come into play. If an animal hurts anybody or wrecks people's property, the victims would have a private cause of action to sue for damages. Strict liability falls to whoever is responsible—whether that is the landowner, the person who created the animal, or, in the case of de-extinction, the owner of an unextinct pet who let it off its leash. So if a recreated mammoth ever broke free from its enclosure and ran around stomping on other animals, destroying cars, wrecking gardens, or who knows what else, its owner would probably be liable for the damage. Species resurrectors and rewilders beware! Then again, this is all early speculation, and laws still need to be made. As Carlin says, "It remains to be seen how the world will react to this inconceivable prospect of de-extinction becoming conceivable."

Much of the legal information I have been able to gather pertains to North American and European law, but Torrance had some interesting ideas to share with me about how de-extinction may function differently in non-Western nations like China, where scientific capacity is marching forward at a tremendous rate and where technology is strongly regarded as an engine for increasing quality of life and wealth. In the West, most conversations about emerging technologies concentrate almost as much on their potential negative consequences as on their possible benefits; in China, however, Torrance suggests, "innovation is almost seen as an unalloyed good, and there is much less skepticism about it than we have here... It is no coincidence that a lot of the splashiest headlines about CRISPR and its applications are coming out of China." The first primates to be edited with CRISPR—twin crab-eating macaques—were engineered in China in 2014 by Xingxu Huang, a geneticist at Nanjing University's Model Animal Research Center, and his colleagues. Then, in 2015, a group of researchers led by Junjiu Huang at Sun Yat-sen University in Guangzhou became the first to publish a paper demonstrating the use of CRISPR-Cas9 in (nonviable) human embryos, stirring up worldwide controversy about what they might be unleashing. Soon afterward, other nations, places like the UK and Sweden, started allowing similar things. In 2016, the first human trial using CRISPR took place, again in China: a live lung cancer patient was injected with gene-edited cells in an undertaking led by oncologist Lu You at Sichuan University. Torrance thinks there is an appetite in China to lead the way with CRISPR because science has been such a positive influence in their economy and the rule of biotechnology law is still at a relatively early stage—compliance with regulations is sometimes a less bureaucratic process there than in the West.

If Torrance is right and a lot of the advances with CRISPR continue to be made first by China, that then leads to the question of whether their actions will drive other countries to move faster when they would otherwise opt for a slower approach. Let's say that China is able to demonstrate increasing successes with CRISPR, in humans or in de-extinction, and maybe even able to bring a couple of creatures back, such as the Yangtze River dolphin (though hopefully not without unpolluted waters for it to live in). Might more precautionary countries start to feel bogged down by their homegrown bureaucratic approach? "I could see that happening," Torrance says. "I could see us feeling like we're being left behind, and thinking, 'Well, gosh, I think we really ought to be a player in this field because if we're not, then they will have a competitive advantage.' This is all speculation, but we don't exist in a vacuum."

On the other hand, he can also see the frontiers of CRISPR being pushed to the edges of human ability inside privately funded laboratories that are outside of the public eye. For example, if rich philanthropists with deeply personal interests in doing de-extinction want to get involved in setting up their own research facilities in the U.S., there's not much that can stop them. Privately funded labs are largely exempted from standard regulations, because those regulations typically pertain to federal funding for research being done with federal oversight. So while one of the hotspots for de-extinction might be a country like China, another center might be secret or not-so-secret privately funded laboratories elsewhere. The more that CRISPR becomes normalized as a mundane laboratory process instead of a revolutionary technology—which is what's happening—the less likely it is that there will be a large regulatory burden associated with using it. And as more trials get underway, the legitimization of CRISPR as a standard laboratory tool will make it increasingly difficult for any sort of omnibus law to be passed outlawing de-extinction.

Reflecting on this, Torrance says, "I think by the time that we can actually do de-extinction, people will just yawn and say, 'Oh yeah, that's just those biotech people doing good stuff, doing crazy stuff... Just let them do their stuff!' " Scientists working with CRISPR today might not know they are aiding de-extinction, but by making the gene-editing tool so easy to use and effectively functioning, they inevitably are. On a grander scale, their research directly influences how many candidate species may actually undergo de-extinction. In this way, the CRISPR boom has a trickle-down effect toward making de-extinction a reality. While some countries are far ahead on this, enough groups work with CRISPR now that in just the last five years, it has already gone from an obscure acronym to a household name for biotechnologists all over the world. I can't begin to imagine where its prevalence will take us by the time the first gene-edited passenger pigeons or woolly mammoths are expected to arrive.

CHAPTER SEVEN

CAN DE-EXTINCTION SAVE SPECIES ON THE BRINK?

A Scientific Setup

IN THE SPRING of 2013, at Clare College in Cambridge, England, a group of conservation biologists found themselves on a blind date with a group of synthetic biologists. At first it seemed the date wouldn't amount to much more than an exercise in awkwardness. It was hard to tell if the participants had anything in common with each other. After all, the conservation biologists were much older than the synthetic biologists, and a bit weary about the state of the world, while the synthetic biologists were bright-eyed and optimistic about life in that zestful way youth often are. They had a bunch of new tools to offer and were a generally hopeful bunch themselves. But even though it wasn't exactly love at first sight, there was a palpable sense that, if the two groups would just spend some time getting to know each other, many beautiful things might be born.

A principal goal of synthetic biology is to make biology easy to engineer. By combining standardized genetic parts in novel ways within a cell, scientists can engineer organisms to produce desirable things, such as biofuels, flavors, fragrances, or drugs, as part of their biological processes. In a way, synthetic biology is a bit like writing software for cells that makes particular cellular programs run. Synthetic biologists also try to understand how biology works by building it themselves from the bottom up or stripping away components to get to the minimum set of biological bits needed to make a system tick (such as by removing all of the unnecessary genes in a microbe's genome until it can just barely keep its functions alive). The field applies an engineer's mindset to the biological world, creating new possibilities for life that nature does not provide on its own. And when they're compared with conservation biologists, who've been fighting extinction for a long time, synthetic biologists are the new kids on the block.

When the two groups met, the idea was that if their date went well, perhaps they would see each other again. And if it went *really* well, maybe they could even conceive little baby endangered animal clones together.

De-extinction came up in their conversation, but so did other things they might like to do, such as reengineer corals that would be able to resist the devastating effects of climate change. Or make synthetic rhino horn in the lab, as a company called Pembient is now doing, to deter people from buying horn-based products that poachers kill rhinos for in the wild. The opportunity felt especially rich for synthetic biologists to help conservation biologists conquer invasive species, such as rats, which are responsible for loads of extinctions on islands. Scientists are good at eradicating them, but the means are often overly destructive—even dropping rat poison by helicopter, and one can well imagine that the impacts of that can be devastating when other native rodents are around. What if the synthetic biologists could create a species-specific poison for the conservationists to use—one that works only in the presence of the targeted rat's genes?

The two groups chatted about white-nose syndrome, which has been decimating North American bats. The condition is caused by a fungus that burrows into bat noses and wings while the bats are hibernating, forcing them to wake up and spend energy they don't have enough of. Eventually this kills them. According to Bat Conservation International, by eating insects and pollinating plants, bats contribute an estimated $23 billion to agricultural and human health worldwide every year. So there is a lot of concern about the large-scale effects of the fungus on our food systems, among other things.

Breeding fungus-resistant genes into the bat population would be an attractive idea if bats had a faster reproductive cycle. But bats produce only one pup a year, so conservationists can't afford to wait for breeding methods to eradicate the problem. If synthetic biologists could design a genetic system that attacks the fungus without stopping the bats from eating insects and pollinating plants, much more than industry profits might be saved.

Despite the conservation biologists' growing excitement at the meeting, they had some concerns about getting into a relationship with synthetic biologists too fast. Every time they've jumped into bed with a risky technology, they said, they've ended up getting screwed. And after all their years of working with invasive species, some were worried about synthetic organisms getting out into the wild and becoming new types of invasive threats. Still, working in conservation hadn't been all fairytales, so some thought that hooking up with young, talented synthetic biologists might help them gain some confidence about their future career prospects. If together they could produce something of value for the world, maybe they'd feel better about what they've achieved.

At the same time, some of the synthetic biologists were intrigued by the older, wiser conservationists and thought they should meet again. Today, the two groups are still getting to know each other better, and determining exactly what they would like to get out of their relationship. What remains to be determined is if this is all an exciting fling, or a long-term partnership with a strong future ahead of it. Personally, I'm rooting for the latter.

Conservation's Cold Storage

THE INACCURATE IDEA that we can bring exact versions of extinct animals back from the grave gets de-extinction into trouble—not only because it gives the public an inaccurate image of the science that grounds it, but also because it obscures other, more clearly beneficial ways that de-extinction might change the world.

For example, researchers are tinkering with many of de-extinction's technologies to try to pull critically endangered species back from the brink of extinction. Archiving is crucial in de-extinction, and the genetic engineers who do it work with some of today's most meticulous librarians. Once the genetic code of a long-gone species is sequenced, it gets stored in databases. Scientists can then later recall the strings of code from the databases according to their research needs. Sometimes they even file the physical strands of an animal's DNA deep in the stacks of an archive that's literally frozen in time. The Frozen Zoo at the San Diego Zoo Safari Park is one such place. It's run by the park's Institute for Conservation Research and contains frozen cells from over ten thousand individual animals and a thousand unique species.

After an endangered animal dies at the San Diego Zoo, employees immediately start to pick the body apart—a chunk of a toe here, a draw of blood there, sometimes a jab at the animal's gonads. But there's not an ounce of menace in this ritual; it's just a different expression of care. By rushing to freeze biological samples of the deceased in liquid nitrogen, they're preserving the animal's DNA and, in some cases, whole reproductive cells. "Each cell of an individual is capable of producing the entire individual," geneticist Oliver Ryder, director of genetics at the Frozen Zoo, explained in an Al Jazeera documentary about the Frozen Zoo. By treating the biopsies they collect right after the animal dies, Ryder and his team are able to freeze them such that they won't be destroyed by the destructive processes that occur inside cells at low temperatures. This freezing puts the cells in a state of suspended animation, pressing *pause* on their biological functions without completely breaking them down. When the cells are thawed later, the biological processes inside them might then be able to be booted back up with some special assistance. This can work—as it did with Celia, the last bucardo, when she was briefly made unextinct with the help of frozen cells. The DNA inside the cells that were taken from her ear and frozen in liquid nitrogen right after she died was the most crucial ingredient for cloning her. But cells at the Frozen Zoo don't just come from extinct species; they also come from threatened populations.

The Frozen Zoo is part of a movement that transforms zoos from outmoded menageries into participants in conservation. Many such programs exist, like the UK's Frozen Ark, which is run by the Zoological Society of London, the University of Nottingham, and the Natural History Museum. In the U.S., the American Museum of Natural History houses eight cryogenic vats for the same purpose in an underground ark of its own. A program with even grander ambitions—Noah's Ark, as it's known—is based in Russia at Moscow State University. It aims to store all of life's biodiversity, not just the DNA from creatures facing the threat of extinction, in liquid nitrogen. These programs provide the first frozen habitats for creatures whose fates may be shaped by our own technological talents. But some people say the programs are defeatist because they suggest that life can be "banked" and resurrected at a more optimal time.

Frozen DNA also creates a rich imaginary vision of what a curated wilderness could look like. In the deep freeze, scraps of species are combined in new ways that would never occur in the wild, as, for example, panda bear cells may sit next to tissue from a giant tortoise. Frozen zoos ultimately create objects that can travel into the future with multiple possibilities for what might happen to them. Freezing cells is about staving off species death, trying to keep populations alive after they have diminished in the wild—and, with the dawn of de-extinction, allow some version of the dead to rise again. The science understands life as something that is perpetually put on hold and potentially restarted, without strict pressure to do so at a specific time. In this sense, the cryopreservation of species has something to say about when—and if—a gene pool is really extinct. But it is also about locating the present in time and picturing what we might see in the future. In this way, the present is a history of the future that brings some big responsibilities with it. It asks questions about which lives are worth saving, which lives are worth re-creating, and, ultimately, which lives are not worth the effort.

There are two main ways that genetic engineering is currently being used to give endangered species another chance at flourishing in the wild. The first is by increasing the diversity of DNA in a population that has become genetically depauperate, or impoverished. A phenomenon known as a *population bottleneck* occurs in a population when its environment changes so rapidly that only some individuals with certain genetically encoded traits are able to survive the shift. Picture a bottle full of different-sized beads. Before the bottle is poured, the beads inside are nestled among each other and the bottle is full. But when the bottle is turned upside down, the beads slam together in a gravity-driven race to escape. This jams the opening, and many beads get trapped inside. The beads that make it through the bottleneck represent the genetic diversity left in a population after an event of drastic environmental change. Such an event can cull the diversity in a population very quickly and make species vulnerable to environmental shifts that may still occur. The thinking here, then, is that by introducing genetic variation into a population after a bottleneck event has taken place, species may be better protected. Conservation biologists have been doing this for ages by mating endangered populations with other populations and subspecies that carry more variation in their gene pools. But genetic engineers, like synthetic biologists, now have more precise tools.

An example of how this may work appears in the Woolly Mammoth Revival. One of the project's stated goals is to assist in the conservation of living elephants, many of which are endangered in the wild. By modifying the elephant genome with woolly mammoth DNA, researchers hope to create cold-tolerant elephants that have greater genetic diversity and that are able to live in habitats in which they otherwise could not. If the project is ever successful, this intervention could expand the endangered elephants' native habitat range.

For an example that works strictly within one species, take the black-footed ferret. The last wild ferrets were wiped out in Wyoming, partly because people deemed prairie dogs—the ferrets' main source of food—to be pests and went overboard in removing them. In 1987, some conservationists managed to collect twelve surviving ferrets, and combined them with six already in captivity to start a breeding program. Only seven of those eighteen ferrets were able to reproduce successfully. Thirty years later, with the help of the U.S. Fish and Wildlife Service, the Smithsonian, and other institutions, there are now about 8,000 black-footed ferrets that all come from those seven founders. Despite this increase and the fact that 4,100 individuals have been reintroduced into twenty sites across eight American states, Mexico, and Canada, the species has still not been properly restored. Partly this is because its wild habitat keeps diminishing. But the animals face another disadvantage independent of that fact: all of today's living black-footed ferrets descended from those seven founders, so their diversity is constrained by a tight bottleneck. They suffer from low fecundity and have problems reproducing on their own. Scientists are now trying to revitalize their genetic profile with a high-tech diversity-boosting approach.

When the U.S. Fish and Wildlife Service caught wind of de-extinction, they approached Revive & Restore to see if their researchers had any ideas that might help them ferret out a solution to this problem. Phelan and Brand were up for the task and started a black-footed ferret recuperation program using genetic rescue techniques. They began by sequencing the genomes from two living ferrets and two ferrets that were in cold storage in San Diego's Frozen Zoo. By comparing the ferret genomes, they were able to see that the population had experienced a serious loss of genetic diversity over time, marked by signs of inbreeding. Phelan and Brand propose that genes from more genetically diverse dead ferrets be introduced into the living population, giving it a beneficial genetic boost.

But that's not the only engineering that Revive & Restore thinks might help. One of the biggest threats to the black-footed ferret is an infectious bacterial disease called *sylvatic plague,* which spreads by way of the same bacteria that cause bubonic plague in humans and is carried by fleas. Sylvatic plague also affects prairie dogs, which ferrets feast on, making it a double whammy of devastation for these guys.

Here arises the second way that the technologies used in de-extinction might help restore endangered species—by creating new individuals resistant to pathogens and parasites that threaten species in the wild. If, for example, Michael Archer and his team ever succeed in cloning the gastric-brooding frog, they could try to introduce chytrid fungus resistance into their re-created frogs' DNA by tweaking their genes. This way, the clones might stand a chance in the face of the fungus that made their predecessors go extinct in the first place. Phelan says a light went on in her head when she realized that they could explore biotechnological approaches to conserving endangered species in tandem with their de-extinction work.

"It would be incredible if we could do this," she says. "So many species, you know—bats with white-nose syndrome and avian malaria that is knocking out all sorts of birds in Hawaii with a non-native mosquito. There are so many challenges for endangered species that it is no longer just about habitat issues. It is also about invasive wildlife diseases." It would be wonderful if researchers could use this technology to clone a genetically diverse member of a species that lived before a bottleneck event, or could isolate beneficial parts of its genome that might increase the variance in a genetically depleted population. It would be even better if researchers could make a species' genes also express disease resistance or could engineer DNA insertions to create the same effect. And since those possibilities dawned on Phelan, she's been thinking hard about how we might dial down wildlife diseases with synthetic biology techniques.

A Biotechnological Boost for the Northern White Rhino

WHEN JOURNALIST M.R. O'Connor set out to write her book *Resurrection Science,* there were seven northern white rhinos left on the planet, the result of rampant poaching. By the time she finished, there were only five. Three of them live at the Ol Pejeta Conservancy in Kenya, where drones, dogs, and armed rangers protect them twenty-four hours a day. For much of the time that I've been writing this book, there have only been four northern white rhinos left, and as I'm typing this, just yesterday, Nola, the forty-one-year-old female living at the San Diego Zoo Safari Park, was put down after she didn't recover well from surgery. That leaves just three northern white rhinos on Earth—all of them living near the equator in the Kenyan conservancy. By the time this book is published and in front of your eyes, there could very well be fewer, though I desperately hope not.

On June 1, 2016, I found myself traveling down a bumpy gravel road in Kenya in the back seat of an old boxy car with my fiancé, Sebastian, our friend Moses, and Solomon, a ranger who works at Ol Pejeta. We cruised slowly past zebras, impalas, warthogs, and elephants that lined the driveway to the endangered animal enclosure, where the black, southern white, and northern white rhinos are kept.

At the enclosure's entrance, Solomon hopped out to get his colleague James Mwemba, who has been working with their northern whites since 2009, when they were flown over from Dvůr Králové Zoo in the Czech Republic. A breeding program in the Czech zoo tried for years to restore the population but wasn't very successful. Only one captured female gave birth there, and only one of her offspring managed to reproduce in the next generation. Eventually, the Czech zoo and the conservancy in Kenya decided it would be best to return the rhinos to their African homeland to see if their native environment might spur them to procreate with more success than they had in Europe.

Mwemba has been through a lot with the world's last three northern white rhinos. Every day that he's on the job he feeds them, rubs them, talks to them, makes sure they're drinking well, looks out for their emotional well-being, and is their first responder if anything goes wrong. Early one morning in October 2014, while making his rounds of the endangered animal enclosure, he noticed that Suni—one of the conservancy's two northern white rhino males at the time—seemed to be sleeping at the extreme end of the enclosure in an unnatural position. Normally at that hour the northern white rhinos are wide awake, so the fact that he wasn't moving concerned Mwemba. As he approached Suni, he could see that his belly was fully distended—filled to the brim with gas. He didn't need a vet to tell him that Suni was dead.

Distressed, Mwemba couldn't quite process what he was seeing. He knew that no poachers had made it onto the site overnight, and just a day earlier Suni had been teasing him while they played together, scrubbing his hind legs on the ground like he always did. He had eaten well and spent a great deal of time in the mud, which, for a rhino, is like a day at the office for many of us—totally expected and part of the drill. The postmortem exam indicated that he had died of natural causes at just thirty-four years of age. Despite the grim circumstances, this relieved Mwemba a bit. He felt better at least knowing that Suni hadn't died from a lack of proper care or protection. But his death raised the stakes for this species, leaving the conservancy's other male, Sudan, as the only sperm-producing northern white rhino in the world. Sudan got his name from the country where he was captured at just two years of age. When that happened, he was taken to the Czech Republic, where he was entered into the zoo's breeding program.

Eventually, he was brought back to the continent from which he first came.

Today at Ol Pejeta a sign outside his enclosure reads, "The three Northern White Rhinos are used to being approached by people. However, all wild animals may be unpredictable," leading me to expect that all three of them would be on the other side of the gate. But when I step inside, I can see only Sudan, way over on the opposite side of the paddock. Mwemba explains that that's because he is now too old to live with the other two northern whites—which, in 2016, are his twenty-six-year-old daughter, Najin, and her fifteen-year-old daughter, Fatu. It is believed that Sudan's independence minimizes the chances that he will get into an accident. They fear that if his female relatives were to playfully push him or if he were to get too excited by their presence and fall over, he might fracture something and die, the way old people can when they break a hip.

These rhinos generally live between forty and fifty years in captivity. Already, at forty-three, Sudan spends his days all alone except for sporadic company from the patrollers who surveil him 24/7 and feed him bananas, carrots, and horse cubes at 4 p.m. as part of his daily routine.

"What's a horse cube?" I ask Mwemba, and he motions for me to accompany him into a shed with one of its doors already slung open, as if to invite me in. Inside the shed are crates upon crates of bright orange carrots, and several burlap sacks filled with brownish-green pellets, which I think look an awful lot like the Canada goose droppings that cover the Toronto Islands in the warmer months. But I understand what they are when Mwemba plunges both of his hands into a sackful. They look extremely fibrous. I pick one up and put it in my pocket, not because I expect to be able to feed it to Sudan, but because I want a souvenir of the place. I had assumed I would be spending the afternoon observing rhinos from afar, but once we leave the shed, Mwemba walks me right up to Sudan, so I can see this prehistoric-looking beauty up close. I am really excited, but also ready for Mwemba to tell me to halt at any second, before I get too near. However, this doesn't happen. As I approach, I see that Sudan is lying in the corner of the paddock under a roof that covers more than half of his enormous cement-colored body and the rectangular pan of water beside him. His head is just poking out from shadow.

"Sudan!" Mwemba yells, and to my astonishment he responds by turning his head like an obedient dog. I'm shocked that he knows his name, and as I'm pondering this, Mwemba starts saying something that sounds like an apology. "Oh, I'm sorry, I'm sorry for that—it's a sign of good health." For half a moment I don't know what he's talking about until I realize that the sound I'm hearing is the beginning of a twelve-second-long rhino fart that sounds like a muffled machine gun, which causes me to take a step back. After it's over, I wait for several moments more, stunned at the decibel level of his digestion, just to make sure the air's clear before I creep in closer.

Once I'm relatively confident it's safe, I slowly crouch down next to Sudan's nearly three-ton body and sidle up to his side, placing my hands gently on his mud-covered neck. Most of the mud has dried and started to crack, and I'm all of a sudden reminded of what I look like when covered with a facial mud mask. I have no words to describe what it's like to touch a creature that you know is the last male of its kind on the planet, and one of the last individuals in the world, except that it's purely humbling. The experience of meeting, observing, and touching Sudan puts me in a reflective state of mind that stays with me for days. It certainly feels like the most important encounter I've ever had with a non-human animal.

When Sudan was first brought over from the Czech zoo, he was in a bigger enclosure, 780 acres, with the two females and some southern white rhinos that the zoo tried to get him to crossbreed with. Although there were matings, no successful pregnancies occurred, perhaps partly because Sudan's sperm count is very low. Even if it were higher and he did manage to mate with the two remaining northern white females, that still wouldn't be a perfect solution: the last living ladies are his daughter and granddaughter, which could create serious inbreeding problems. But making an inbred calf with either of them wouldn't work out anyhow, for the females have reproductive problems of their own that make full-term pregnancy pretty much impossible. There once was hope that a northern white rhino calf would be born at Ol Pejeta before Suni died because there were records that he had mated with Najin. But after all kinds of inspections, no pregnancy was ever detected. Although at one point Najin was able to give birth, she now has a problem with her hind limbs that make her unable to support the weight of another pregnancy. And her daughter, Fatu, the youngest northern white rhino alive, is infertile as a result of a disorder that prevents embryos from implanting in her uterus. Since the last three rhinos are fragile, old, or infertile, it would seem that the species faces an evolutionary dead end.

In December 2015, a group of researchers met in Vienna to discuss what might be done about this crisis and came up with some high-tech plans for saving the northern white rhino. The first idea was to perform in vitro fertilization (IVF) by combining sperm from Sudan, or from any of the four males whose sperm is currently stored at the Frozen Zoo, with frozen oocytes (immature eggs) or with oocytes from the living females, and then implant the embryos in a surrogate mother. But since Sudan is the father and grandfather of the last remaining females, his sperm was ruled out. Frozen sperm could be used, but there are no frozen northern white rhino oocytes available. Scientists would have to get them fresh from Najin or Fatu, a process that requires anesthetic and a specialized lab, but no lab close to where they live in Kenya currently does this. Processes for IVF will need to be developed and tailored to the location, then tested on the less scarce southern white rhinos to make sure they are working efficiently before researchers try them with their much rarer relatives. Even then, without eggs from deceased northern white females available, any new rhinos made this way would be created from the eggs of only two females and the frozen sperm of four males, ensuring that they would have extremely low genetic diversity. To increase their odds of producing healthy calves, the researchers need to start somewhere else.

Another proposed solution is to save the northern whites with the same technology that the Woolly Mammoth Revival project is using to reprogram adult skin cells—fibroblasts—into induced pluripotent stem cells. Remember, these are cells that have been transformed to a highly undifferentiated state and can then be coaxed into becoming nearly any type of cell in the body.

Scientists have already created stem cells this way using cells from Fatu's skin. The next step is to prod those stem cells into becoming specialized sperm and egg cells that can be combined—in vitro—to make northern white rhino embryos. This technique has been reported to work in mice but not yet in rhinos. If it does work, the embryos will be implanted in a surrogate mother with the hope that she can bring one to term. But given both the rarity of northern white females and their fertility issues, a surrogate from the southern white rhinos would be a better choice. Whether southern and northern white rhinos are entirely separate species or are subspecies is still disputed, but either way, several southern white females already living at the San Diego Zoo Safari Park and Ol Pejeta Conservancy might work well for the task.

But making this transfer is not without its problems, since the rhinoceros has a highly convoluted cervix that is extremely difficult to penetrate. It has been suggested that a horse could be used instead, which would make for an unbelievably surreal birth. But if researchers eventually get surrogates to birth northern white rhinos, they'll be able to restore their population through three methods that can work simultaneously: (1) old-fashioned mating within the new generation of northern white rhinos, (2) mixing of sperm and oocytes in vitro to create embryos that get implanted in a surrogate mother, and (3) production of stem cells that can give rise to sperm and eggs that combine in vitro to make embryos to be implanted in a surrogate mother. Even using all three methods, researchers estimate that it will take at least fifty years for the northern white rhino population to be restored to non-endangered status.

This project faces many of the same ethical issues as other de-extinction endeavors. Several fertilized embryos might not take, and newborns could die before they leave the lab. No one knows how a northern white rhino calf's microbiome or hormone interactions with its mom will be affected by being born and raised by a surrogate mother that's not from its own exact species. For example, if the southern white rhinos' milk is not suitable for newborn northern white rhinos, would they have to be hand-raised by humans? And if so, how would that affect not only the baby but also the southern white rhino surrogate mother that grew and birthed it?

Uncertainty about whether the northern white rhino is a separate species or a subspecies along with the more abundant southern white rhino raises another interesting ethical question. The researchers write, "Is rescuing a species or a subspecies important enough to justify subjecting members of another species or subspecies to medical interventions such as ovum pick up or embryo transfer?" In other words, is it ethically okay to subject the southern white rhino to invasive medical procedures just because the northern white rhino population is in perilous danger? And does the answer change depending on whether or not the two types of rhino are subspecies or separate species? Who gets to decide whether or not that sort of difference matters?

When I ask Mwemba what he thinks of these questions, he doesn't seem troubled by them. "I personally believe in science, and I believe in the artificial methods to bring these species again to regeneration," he says. "I am wholly optimistic that it is going to work, because I know there are great minds behind this." After having met Sudan, Najin, and Fatu, I'm hoping that he's right.

Gene Drives

SEVERAL SCIENTISTS I'VE spoken with are excited that a powerful and controversial tool known as *gene drive* might revolutionize the biotechnological fight against wildlife diseases. Although it has received lots of attention in recent years, the idea of gene drive is not new. Scientists have been aware for decades that it occurs naturally in the wild, but its power has been tricky to harness. What's new is that tools like CRISPR now enable scientists to take advantage of gene drives in more precise ways to alter nearly any gene in a sexually reproducing species that will propagate it.

Here's what's cool about it: in sexually reproducing creatures, most genes have a 50 percent chance of being inherited in the new generation. But gene drive promotes the inheritance of a particular gene to ensure that it gets passed on, and thereby promotes that gene's prevalence in a population—even if the gene is harmful to the organism. The gene that's "driven" into a population could make individuals sterile, for example, and therefore could wipe out a rapidly reproducing population in just a few short generations. Or it could even make an organism resistant to a disease-carrying parasite, which would mean the organism is no longer a vector for that disease, preventing its spread to other hosts. A classic example of how one might use gene drive technology is to engineer mosquitos with CRISPR so that they no longer are susceptible to the parasite that spreads malaria to human populations, and so that they carry the genes required to propagate that protective trait in the next generation, and every other subsequent generation.

Another way of looking at it is to say that gene drive makes the genetic engineer—the scientist in the lab—obsolete after the first generation of organisms is engineered. Using the mosquito example, if a genetic engineer wanted to make sure that all generations of that mosquito species would express a gene that makes it resistant to malaria-carrying parasites, they would have to engineer the first generation of parasite-resistant mosquitos and let them out in the wild. Once the mosquitos are out in the wild, they will meet wild-type, non-engineered mosquitos, and eventually make babies with them. But those babies will only have a 50 percent chance of carrying the parasite-resistant gene that the human engineer put into the first generation of lab-made mosquitos. And *those* babies' babies would then only have a 25 percent chance of inheriting the gene, and so on. The probabilities of inheritance are therefore working against what the engineer wants, and that engineer will have to do more genetic engineering by hand if he or she wants to produce more mosquitoes with the parasite-resistant gene. But with gene drive, once the genetically engineered mosquitos make it into the wild and mate with wild-type mosquitos, they will produce baby mosquitos that inherit the parasite-resistant gene nearly always. And that new generation's babies will also carry the parasite-resistant gene, because that's the gene that's being "driven" into the population. In theory, this allows the genetic engineer to spend time doing other things.

Allow me to paraphrase the excellent explanation of gene drive from the fantastic radio show *Radiolab*: When the genetic engineer is working on editing the gene that causes some form of malaria-carrying parasite resistance into the first generation of mosquitos, they make an additional edit right next to where they made the first one in the mosquito genome. This second crucial edit encodes the genes that are needed for the CRISPR system to work in the mosquito on its own and tell it to make that particular protective genetic tweak in the next generation. So when the engineered mosquito that has both the protective anti-parasite gene and the genes needed for the CRISPR system meets an unengineered mosquito in the wild and makes babies with it, those babies end up having two sets of genes inside of it: one from mom and one from dad. However, the set of genes that came from the engineered parent mosquito have this pair of CRISPR scissors encoded into it that get expressed and that can travel over to the other set of genes from the wild-type parent. The scissors will edit that set of genes the baby got from its unengineered wildtype parent so that they end up looking identical to the ones that the baby got from its engineered parent. Now the baby mosquito has two sets of the parasite-resistant genes in its genome, which means it will express that protective trait. As Jad Abumrad, cohost of *Radiolab*, described it, it's a bit like "allowing that mosquito parent to pass the scissors to the baby, and snip snip snip, and then that baby passes the scissors to the next baby, snip snip snip, and it is literally like a chain reaction." It's as though the successive generations, racing down the line, are passing a baton forward in a relay race.

Using gene drive, scientists at the University of California, Irvine, did this in the *Anopheles stephensi* mosquito, a primary vector of malaria that infects human populations. Two genetic mutations were engineered into the mosquitos, the first of which stimulates the production of antibodies right after a female mosquito has a blood meal. The antibodies then attack and destroy the malaria parasite if it is present. The second mutation ensures that the gene drive self-perpetuates in the next generation, which happened with 99.5 percent efficiency. The journal *Nature* describes it as "a gene that could spread through a wild population like wildfire."

Although people are excited about gene drive's mitigating properties for human health, the technique also holds great promise for wildlife conservation. What if scientists could use gene drive in flea populations that carry sylvatic plague so that the fleas are no longer susceptible to the bacteria that transmits the plague? That surely would help with the genetic rescue of black-footed ferrets, which easily contract sylvatic plague because they eat prairie dogs that fleas regularly infect with the disease. This works well with fast-reproducing creatures as a genetic rescue technique. However, if a gene drive were introduced into a species with a long reproductive life cycle, like humans, it would take hundreds of years before any effects were seen.

Gene drive may sound like a solid solution to wiping out vector-borne diseases and even some invasive species, but it could have serious adverse effects. For starters, once a gene is driven it into a population, its effects could be hard or even impossible to reverse. What if the gene that is programmed to be inherited with almost absolute certainty somehow spread beyond its bounds, wiping out gene pools that it wasn't meant to?

Stewart Brand understands that the idea of gene drives may trouble those who are already critical of using biotechnology to craft aspects of the natural world. Referring to the birds of the Hawaiian archipelago that were rendered extinct by avian malaria, he tells me that bringing those birds back now might require intervention beyond the techniques used to get them back, creating an entirely new crop of ethical issues. "Well, you're mucking around with the genes of these animals anyway," he tells me, "and if you can bring back the Hawaiian 'ō'ō and we can tweak it so that it's not susceptible to avian malaria, is that okay? Personally, I think it is okay, but I can see how lots of people would say, 'Now you really are playing god. If it wasn't there in the original bird, you aren't supposed to add it.' " If it were up to him, he would "tweak the damn birds," but he recognizes that the uncertainties about unintended effects require a great deal of research before the technique is used.

Tweaking the American Chestnut

ONE GOOD EXAMPLE of protecting an organism from infection by tweaking its genes comes not from the animal kingdom, but from the wonderful world of plants. Back when the Europeans arrived in America, the American chestnut was the most abundant tree in the east, making up one out of every four trees in the region and looming more than 100 feet above the rest of forest life. The American chestnut was a keystone species with a stable nut crop that fed a slew of forest dwellers, including passenger pigeons, blue jays, deer, bears, and raccoons. Humans, of course, liked the nuts too, which a certain Christmas song describes. We also liked to build furniture and decks with the chestnut's rot-resistant wood, but that's not why an estimated 4 billion trees from Maine to Mississippi became sickly in under fifty years.

An infection appeared in the 1870s, when we started importing the American chestnut's cross-Pacific counterpart, the Asian chestnut, from Japan to North America. Hidden in the Asian chestnut was a fungus—blight—that had not done obvious damage to the tree in its native range. But it turned out that the American variety was unexpectedly devastatingly vulnerable to the chestnut blight. The blight was discovered in New York in 1904 when a scientist named Hermann Merkel noticed that a tree at the Bronx Zoo was dying from it. But by then, it was already too late to stop its spread. The blight blazed through forests of endemic trees in the northeastern U.S., leaving only dead and dying stems in its wake.

The blight is so deadly because it releases oxalic acid, which attacks the chestnut's cambium (a cylindrical layer of tissue in the stems and roots of many plants) so that a canker forms on its trunk. Once the canker wraps around the circumference of the tree, it chokes the tree's ability to carry water and nutrients between its roots and its branches. As a result, everything above the demarcation line dies. The stump below the canker can still send up new shoots, but it is only a matter of time until a new canker forms. Although millions of sprouts are still out in the forests and it might take several hundred, even up to one hundred thousand, years before the species completely disappears, the American chestnut is considered effectively extinct.

For the last century, scientists have been working hard to postpone the inevitable. They've been spraying trees with fungicides and infecting them with viruses that they hope will attack the blight. They've been exposing the trees to sulfur fumes and making them undergo radiation, but somehow the blight always bounces back. For the last few years, scientists at the SUNY College of Environmental Science and Forestry in Syracuse, New York, have taken a double-pronged approach to researching blight-resistant chestnut trees. At first, they tried crossing the American chestnut with the Asian chestnut in hopes of spreading fungal resistance from the latter's genome into the former's. Although some of the resulting hybrids ended up with blight resistance, they also carried the undesired traits of the smaller Asian orchard tree. A better solution would have been to introduce specific sequences into the American chestnut, one gene at a time.

When scientists set out to try that, no one had mapped the genome of the Asian chestnut tree, so they didn't know which genes would be best to change. But since they wanted genes for fungal resistance, they looked at other, more thoroughly studied plants to see if any had been mapped for that trait. Fortunately, it had already been discovered that wheat fights off fungus with the help of genetic weaponry that makes enzymes that can destroy oxalic acid. So the researchers inserted genes coding for the anti-acid enzymes from wheat into American chestnut embryos. Eventually they got trees that can heal their own cankers by disarming the acid without killing the fungus itself.

On April 18, 2012, the transgenic American chestnut was brought back to the area where the blight was discovered in 1904, when SUNY researchers planted the experimental species at a test site in the New York Botanical Garden. In the future, they would like to cross their transgenic trees with trees in the wild to try to spread its acid-disabling abilities. Former mine sites where trees have been cleared and the soil is unnaturally acidic might be a good place to start. Government approval for environmental release is currently being sought from the U.S. Department of Agriculture, the Environmental Protection Agency, and the Food and Drug Administration. Once they get it, they'll need a lot of help from the public to get the tree's numbers back to over a billion. It's a good thing the American Chestnut Foundation has over six thousand members—their green thumbs will be needed to plant the next generation of genetically modified chestnut trees.

Societal Acceptance of Saving Species with Biotech

OVER THE YEARS that I've been following this research, I've noticed that genetic rescue techniques for endangered species tend to receive more public approval than do straight de-extinction pursuits. Beth Shapiro's take on the merits of the research underscore this idea. In an interview she once said, "The priority of this technology isn't, in anybody's mind, to bring an extinct species back to life. It's to save species and ecosystems that are alive today from becoming extinct." Indeed, Ryan Phelan told me to think about the beneficial domino effects of their genetic rescue work rather than to think about de-extinction alone, and she pointed me to an example that explains why.

When Revive & Restore first looked into Asian elephant biology as part of its woolly mammoth research, the researchers were surprised to learn that a strain of herpes—called *elephant endotheliotropic herpesvirus,* or EEHV—is a major cause of death in young calves, both in captivity and in the wild. The virus usually kills infant elephants up to the age of four within a few days to a week of onset, though there can be a long latency period between infection and the expression of the virus. Once it gets into an elephant's bloodstream, it starts breaking blood vessels and causing its organs to bleed until the hemorrhaging turns fatal. And sadly, once signs of the virus are noticeable to humans, it is already too late for the elephant.

To diagnose EEHV, fluid that has passed through an elephant's trunk is regularly collected and analyzed. If the elephant has herpes, the viral DNA will show up in the trunk wash, and if it is discovered early enough, it can even be treated with antiviral drugs. But treatment is not a cure, and there is no such thing yet for this virus. That's partly because no one has ever been able to synthesize its DNA in a lab to study how the virus makes elephants vulnerable to it. Some researchers are trying to do this now, however. In April 2015, Revive & Restore hosted a conference that gathered fifty-two specialists to brainstorm genetic rescue techniques that could target wildlife diseases and invasive species. Paul Ling, a leading researcher studying the virus at

Baylor College of Medicine, was there and met George Church. Their shared interest in proboscideans came up in conversation, and before long, they had assembled a team to try to synthesize EEHV in Church's lab.

Researchers make viruses in labs all the time—for example, the polio virus was synthesized in 2002, and the 1918 flu virus was synthesized in 2005. Doing so with EEHV might allow researchers to develop a system in the Asian elephant genome that makes elephants resistant to EEHV altogether. Phelan's eyes widened when she told me this. "Can you imagine how profound that would be if we could have more examples like that? Where we know there is a disease susceptibility and the solution for it could be as simple as altering a few genes?"

Ling's group had already sequenced the virus from pieces of DNA they found in trunk washes from infected elephants and gave its genomic sequence to Church so that his team could attempt to synthesize it in the lab. When that's done, they hope to transfer the virus into bacterial cultures that can grow many copies of it and then insert those viral copies into Asian elephant cells to test what they do to elephant tissues. This will allow them to discover preventive vaccines and treatments for EEHV infection by searching for the proteins that the virus uses to infect its host and then creating synthetic systems that attack those proteins. In humans, for example, a certain group of proteins is produced before the herpes virus forms, while other proteins are produced later. If the same is true for elephants, then a clever strategy may involve halting the development of early-stage proteins by disabling the genes that code for them before any proteins are made. Those genes could be disabled by using CRISPR to cut the essential genes of the EEHV virus that code for those proteins, without cutting DNA sequences in the elephant genome. That way, the viral genes cannot be expressed and the viral particles are not formed. Bobby Dhadwar, who works on this project as well as on the Woolly Mammoth Revival at Harvard Medical School, says, "If we are able to come up with a treatment, that would help people realize that this is not just about doing crazy things with the mammoth. It is about doing elephant conservation as well."

I wonder if there will be any public backlash against saving species with things like gene editing and stem cells as these technologies become increasingly applied. What will happen when the three remaining northern white rhinos die? Will people feel the same about using biotechnology to recreate the rhinos after they're all dead as they felt about using biotechnology to prevent their extinction up until the moment before the last ones expired? In other words, is it somehow more reasonable to be opposed to the idea of de-extinction of the northern white rhino after Sudan, Najin, and Fatu die than to oppose the northern white rhino's restoration now while those three are still with us, even though the same types of advanced biotechnologies are used in both scenarios? What are the material, ethical, and environmental distinctions here between de-extinction and genetic rescue of an endangered species, other than the existence of a few remaining individuals hanging on for dear life?

Would waiting one day after the last individual bites the dust be too soon for the establishment of a de-extinction project to be considered acceptable? How about a year? A decade? A century?

The more I think about this, the more questions I have. Such as: Would it be more or less ethical to put resources now toward resurrecting a species of rhino that's already gone, like the West African black rhinoceros, instead of toward trying to help the northern whites? And even more puzzling: Is the moment when the last individual of a species dies even the real extinction event, or is it impossible to attach the completion of an extinction to a single point in time? What is society's threshold for human intervention along the extinction continuum? Can it even be said that there is one threshold, or might society have many, depending on the specific candidate species being considered? We seem to widely recognize that endangered species have a right to be here. But when Phelan asks me, "If they have a right to be here today, why don't they have a right to be here tomorrow?"—meaning the day after the last member of a species has died—I'm left scratching my head.

CHAPTER EIGHT

IS SOME KNOWLEDGE TOO DANGEROUS?

*Learn from me . . . at least by my example, how
dangerous is the acquirement of knowledge . . .*
—MARY SHELLEY, *Frankenstein*

THE WAY WE tell stories about science matters, and though
de-extinction sounds cutting-edge, its underpinning narrative
is much older than its modern manifestation. At least as far
back as the Bible, we find multiple tales devoted to the idea of
undoing death.

Consider the resurrection tale about Lazarus of Bethany, a
sick man Jesus is asked to try to bring back to health. By the time
Jesus arrives, Lazarus is dead and has been lying in a tomb for
four days. Jesus then says, "I am the resurrection, and the life:
he that believeth in me, though he were dead, yet shall he live,"
and calls out to Lazarus. Miraculously, a man wrapped in cloths
comes forth from his grave in a defiance of death still meaningful to today's world, as can be seen in our use of his name.

Geologists call a group of organisms that were once thought to be extinct but suddenly reappear later in the fossil record a *Lazarus taxon*. Michael Archer has named his team's attempt to de-extinct the gastric-brooding frog the Lazarus Project, and in a media appearance even quotes from the Bible himself: "If we destroyed part of Eden, we are responsible for fixing up that garden. This idea of restoring things that have become extinct has actually a biblical sanction. 1 Corinthians 15 verse 26 says, 'The last enemy that shall be destroyed is death,' " he asserts while standing between dry shrubs and shelves of sandy rock. Looking directly into the camera with his head slightly cocked to the side and an air of excitement about him, he signs off starkly with, "I'm Mike Archer, a paleontologist and species revivalist."

Scholars have shown that the public sometimes comes to understand emerging science more through its depiction in entertainment than through explanations from actual scientists. In this way, when we analyze the risk of new technologies for society, public beliefs about risk and how to handle it may be shaped by show business. Stories can influence policy decisions that will eventually regulate science itself. So our stories, even the fictional ones, can carry weight in the real world.

Sometimes our stories lead us to imagine those who do high-tech science as lone geniuses who eat, sleep, and breathe their work. In our minds, they live in labs lined with gizmos that look as alien to most of us as the surface of Mars. Just as Dr. Frankenstein hid his monster in secrecy, we sometimes suspect that dubious experiments are happening behind the closed doors of testing chambers the public doesn't even know exist. But science today is far more social than these stories suggest. In practice, scientists regularly work with scores of other researchers to discover the minutiae of what makes the world around us tick. At times they even work in public ... quite literally, which can get them into trouble. Let me paint a picture for you of what I mean—a famous one, at that.

In the eighteenth century it was common for scientists to take their work on the road, amusing audiences with their special knowledge. In the painting *An Experiment on a Bird in the Air Pump,* made in 1768 by the English artist Joseph Wright of Derby, a scientist performs in public, acting like a vainglorious villain. He is a traveling scientist, and he shows a group of young people how a vacuum works by sucking the air out of a flask that contains a live cockatoo. Surely, if he continues with his experiment, the bird is going to die, which the onlookers seem to understand from the concerned looks on some of their faces. But the way the scientist exhibits the bird makes it clear that he intends to go on. Does the scientist not have empathy with this living creature or the anguish of his audience? Does the fact that he *can* push further with the experiment mean that he *should*? Does he even recognize his responsibility here?

This painting sets the scientist apart from society. He does not seem to value human attachment and ignores public interest as if it stands in the way of his work. He seeks to thrill his audience by riding that tension between the unknown outcomes of his actions and his scientific prowess. In other words, he is playing with what the environmental and social historian Sandra Swart calls "dangerous knowledge." Now that scientists have the tools for de-extinction at their disposal, should they use them? Or are critics too quick to fret about de-extinction, judging it unfairly through the "dangerous knowledge" trope before letting de-extinction speak fairly for itself?

The "dangerous knowledge" theme has been a long-running thread in stories that warn us about our own hubris. Most familiar are the creation stories, leaving little wonder about why the "playing god" critique so often appears in genetic engineering and species resurrection debates. As Swart writes, "Creation stories matter to people. They are stories about power—a power predicated on knowledge... You steal fire from the gods, you are chained to a rock and your liver gets pecked by avenging avians for eternity. You fly too near the sun, you fall into the sea and drown. You insist on finding out your parentage, discover inadvertent incest and have to blind your own eyes. You eat an apple from the Tree of Knowledge, you are evicted from paradise. You make a deal with Mephistopheles to acquire all worldly knowledge and are doomed to everlasting hellfire. You create life and that creature kills your nearest and dearest—or you. Perhaps it is best summarized thus: you screw with the Natural Order—and you are *screwed*."

Some fear that de-extinction will screw us and other creatures, especially if it is just a vanity project, flexing the muscles of our own talents with technology more than doing any real good. The questions that de-extinction raises for conservation, animal welfare, ecological function, ethics, patenting issues, disease, invasive species, other interspecies ecosystem dynamics, effects on human communities, international regulation, and more are still far from being well understood. But what is understood is the cultural foundation that creation stories have built, upon which de-extinction now stands. Although these stories rightly matter to broad societal imagination, the way we use them in our own lives might matter even more.

Frankenstein—the most famous of the "dangerous knowledge" fictions—explores the unintended consequences of our human science when we relinquish responsibility for what we create. The full title of the book is *Frankenstein, or The Modern Prometheus*. In Greek mythology, Prometheus taught humans how to make fire, which he stole from the gods, and was punished for it by Zeus, who sent birds to eat his entrails. The myth underscores the idea that there are some things that only God should know. Isaac Asimov, the science fiction writer, aptly coined the term "Frankenstein complex," meaning the fear of encroaching on God's terrain through technology. As a case in point, when we fear genetically modified foods, we call them "Frankenfoods."

In 1818, after seeing science demonstrations like the one depicted in *An Experiment on a Bird in the Air Pump,* Mary Shelley published *Frankenstein* and thus planted the first seed of this fear. In those days she was particularly drawn to experiments that involved galvanism, in which scientists animated the dead with electricity. Watching from the audience, she'd study how the scientists attached electrodes to a pair of floppy frog legs and was enthralled when they'd twitch with an electric quiver. Back then, such demonstrations created a sense that innovative research was focused on macabre experimentation at the boundaries of human knowledge. It allowed mere mortals to bring life to the dead and come up with plotlines for God-fearing folklore.

Swart argues that, so far, de-extinction has been predominantly understood through the Frankenstein myth. Swart compares dreams of cloning woolly mammoths to the final pages of Shelley's masterpiece, since scientists are once again hunting their monsters in the far North. The Frankenstein myth even makes its way into the most popular de-extinction story there is: *Jurassic Park.* That story scared us half to death with the velociraptor's stare that had the same "yellow, watery, but speculative eyes" as Frankenstein's monster. By framing de-extinction in this way, we reinforce the sense that scientists not only will make creatures with harrowing consequences for the world but will also fail to heed our concerns.

If you've read *Jurassic Park* or seen the film, you'll know that the amusement park's mastermind, John Hammond, was able to fill Jurassic Park with unextinct dinosaurs only by way of his own deep pockets. But the plan was to make his investment back in multiples from the park's admission prices.

Similarly, in 1999, the Australian Museum in Sydney announced a plan to bring the extinct Tasmanian tiger back to life. Although the plan never materialized, the museum sold the project on the idea that it would position the museum "within the crowded market by creating joint promotions with corporate sponsors." De-extinction for profit—a theme in *Jurassic Park*—is what Swart calls a form of "neoliberal necromancy."

Having knowledge of these stories is dangerous itself, Swart warns, because "mythic narrative strength may obscure a more nuanced popular understanding of a [de-extinction] project." As a case in point, we don't usually recognize the real moral of the Frankenstein story. We don't tend to remember that Frankenstein was the name not of the monster but of the doctor who created him. In the more than thirty film adaptations that have been made of *Frankenstein*, audiences always see the monster's malevolence, but in the novel the monster only becomes evil later on. "Remember that I am thy creature," it moans. "I ought to be thy Adam; but I am rather the fallen angel, whom thou drivest from joy for no misdeed." The monster transforms into something uncontrollable only after Dr. Frankenstein abandons it, horrified at what he's created, and refuses to deal with the consequences of his own actions. The monster even protests his misfortune and says, "I was benevolent and good; misery made me a fiend. Make me happy, and I shall again be virtuous." So the story does not actually suggest that there are some things only God should know. Rather, it shows us why we have to take responsibility for what we have created.

What does it mean to take responsibility for our creations when they're capable of sensing their own existence, as Frankenstein's monster did? And what does it mean to take responsibility for the environment they will eventually interact with? Environmentalism, which takes precaution, and postenvironmentalism, which espouses action, are in a tug-of-war at the center of the de-extinction debate. As soon as the proactive side yanks with extra force, our Frankenstein complex—the fear of unleashing a malevolent monster into the world—kicks in. But again, this misses the point that *Frankenstein* has to offer us.

The anthropologist and philosopher of science Bruno Latour argues that the real lesson *Frankenstein* has to offer is that we must love and care for our monsters so that they do not turn mean. But he does not imply that we should avoid creating them altogether. For example, it is not that Ben Novak should not create unextinct passenger pigeons simply because they are "unnatural" creatures made by human hands, but that he should be careful to show them the utmost of care, love, and concern as he responsibly oversees their entry into—and sustainable living in—this world.

In de-extinction, no doubt, love is there for some from the outset. Novak speaks dramatically about his own passion for the passenger pigeon, and even sounds lovey-dovey when he tells me, "I don't think anyone gets into these projects looking at graphs or looking at data and that's all you see. I mean for me, I know when I look at DNA sequences I see living animals. I see a pair of birds coming together and breeding to make some new offspring. I see these events in history that shape what this data actually looks like ... It's an inspiring topic, and it's a very beautiful thing to be doing. *Art* is the only thing that can sum that up."

For Novak, de-extinction disrupts a sense of time and permanence while teaching him about how to care for life in the present. He says he's discovered a politics of care tied up in his de-extinction work, which entangles him in new caring bonds with life forms across a continuum—from the long gone to the potential. At the same time, others have argued that de-extinction forces us to face tragic questions about the ethical conundrums of caring too much. As a society, we care about mass extinctions, the conservation of wildlands, and our duty to intervene appropriately, but an excessive display of concern says something about the distress we feel deep down. In order to care, and to learn how to best act on that desire to care, we'll have to get our hands dirty. But soiled hands can leave stains that we wish weren't there.

That reminds me of an idea I once came across in the writing of Donna Haraway, an esteemed interdisciplinary scholar of biology and culture, and theorist of "cyborg politics," which neither fetishizes nor abhors technology's ability to change what life is made of but always questions its specific ethical entanglements. She tells us that learning from animals is never an innocent act when humans and other species meet. De-extinction relies on various amalgamations of living species being used as gestational surrogates, gamete donors, rearing parents, and adoptive families for species that don't yet exist, incurring significant costs to their well-being and freedom to live out their own ways of life. That's without even counting the direct costs that individual animals being made through de-extinction might suffer, including paying the ultimate price. Then, of course, should things ever go wrong upon reintroducing an unextinct species in the wild, many more animals already living in the release areas might also come under stress. We are not innocent when we decide to take these risks, and we cannot claim that we didn't know otherwise should things not turn out as planned.

I was curious to know what Donna Haraway thinks about de-extinction, so I asked her for an interview, but she declined because of fatigue from being "chatted out." However, in the email she sent turning down my invitation, she told me, "De-extinction is not my favorite conservation approach, I have to say! In my opinion, biotechnology has a big role to play in taking care of the earth and its critters, perhaps including, in specific cases, some genetic alterations for adaptations to rapidly changing conditions. But de-extinction seems hopelessly rooted in barely secularized creation and salvation narratives and approaches to science that are wrong-headed and wrong-souled, plus rarely recognized for their cultural specificity by the folks who support and do the projects, even as they speak of Biblical themes and characters (e.g., Noah's Ark) and treat DNA as a god-equivalent. I want all that talent and energy to go for other sorts of ecological recuperation—much less resurrection mythology and much more mundane care."

IN OCTOBER 2014, on a stage at London's Serpentine Galleries, a purple-gray backdrop displayed an oversized hand with its index finger pointing at a stuffed garbage bag surrounded by black lines that made it seem like volcanic energy was about to explode from inside the sack of trash. Soft fabric clouds floated above the black lines with words scrawled on them like "cloudy," "remainder," and "turning heavy." The enlarged index finger was actually poking the garbage bag, creating a little hole in its side. While I was peering at the clouds trying to decipher their meaning, my gaze was suddenly interrupted by a slender brunette sitting on stage who lifted her head and started to speak:

"My talk is going to be quite literal about the extinction of species. I stay up at night worried about this. Maybe to some of you that's a normal thing to do." The woman is Jennifer Jacquet, an environmental social scientist and professor at New York University. She is one of several speakers invited by the Serpentine Galleries to deliver a talk at this Extinction Marathon, as it's called—where artists, writers, scientists, filmmakers, choreographers, theorists, and musicians share their visions for the sixth mass extinction. Jacquet says she is worried about the 870 or so species that have gone extinct since the sixteenth century that we know about, species that fell victim to what she calls "conspicuous extinction"—species that we literally watch fade away. She continues: "One of the species that I stay up worrying about and mostly lamenting—I know I'm not alone on this because a friend of mine is torn up that there are no longer giant sloths around—is the Steller's sea cow, which went extinct in 1768 after twenty-seven years of having been discovered by the Russians." She stays up at night from worry, and her friend is in the throes of mourning. This is much more than an intellectual recognition of their absence—they *feel* the fact that these creatures are no longer here.

"I even spoke to a woman who studied survivor guilt and asked her if she has many patients who complain that the human species is surviving while all these other species go extinct. She said it was not a presenting symptom. Maybe I'm a little strange," she adds. Survivor's guilt is a mental condition that people suffer when they feel they have done something wrong by making it to the other side of a traumatic event while others are not so lucky. It is common among those who endure a natural disaster, epidemic, or war, but the extinction of a nonhuman species is a more puzzling case. Although some symptoms of survivor's guilt don't seem likely when it comes to extinction—mood swings and social withdrawal—others, like depression and, as Jacquet claims, sleep disturbance are more common. If enough creatures are wiped out and the natural food chain collapses, our own species' domino will fall with the rest. That might be why she says she boils her feelings down to a misplaced fear about our own human mortality. And yet, it doesn't really feel like we're going to die all that soon, does it? Many of us expect to have children who will be able to have their own children, and so on. How crucial is it for our own survival that we care about extinction, for all species, with immediate and blazing urgency? As cold as that sounds, there are things to discover under the rocks that a question like that upturns.

Gregory Kaebnick, a scholar who researches how biotechnology puts human values at stake, takes that question seriously in his book *Humans in Nature*. He asks why we should feel anything at all when a species is lost, and to find an answer, he refers to the work of the microbiologist Lee Silver. Silver once traveled to Peru to see the last remaining sea otters in the region. He wanted to be one of the last humans to see them alive, but he wasn't quite sure what was pushing him to make the effort. He kept a diary of reflections on the trip. In one of his entries, he wrote about a discussion he had with the group he was traveling with: "I decided to challenge the group with an impertinent question: 'Why should we care if the giant otter species goes extinct?' I saw a silent look of horror on everyone's face. How could I pose such a question, they wondered, to this group of people in this place? Attempts were made at providing rational explanations, but none were truly compelling. Like many other widely shared attitudes toward Mother Nature, the idea that we should care just feels right, although people don't quite know why."

Kaebnick offers thoughts about why that might be: We humans carry moral reactions around with us as though they are hardcore truths. The fact that we hold them as self-evident truths is why it is difficult to offer a rationale for them. Our certainty is rooted in *feeling,* and people just *seem to know* what is right or wrong. Kaebnick says this is because moral stances are always, in part, emotional phenomena and that we should come to understand them that way. "The attention to this topic in environmental philosophy," he says, "is due to the fact that moral concern about the environment is widespread; the idea that the environment deserves moral concern has acquired the status of a settled judgment"—hence the silent horror on everyone's face when Silver questioned it—"and the problem is how to explain it."

But Kaebnick also says, "A purely emotional approach to morality would also be a contradiction in terms; to engage in moral deliberation is to step back from one's immediate reaction and think critically about it. If we relied unquestioningly on our initial feelings about a moral problem, much that we count as moral progress would not have occurred." It is now, more than ever, he suggests, time to consider how we feel about losing species and ask ourselves what moral progress we can make, both personally and collectively, in our time of mass extinction. It's a question that's been haunting me for years, especially as it relates to de-extinction.

I turned to Thomas van Dooren, a leading philosopher in the subfield of extinction studies and senior lecturer at the University of New South Wales in Australia, to help me think it through. I was wondering what extinction can teach us and what lessons are worth taking seriously when de-extinction is no longer the stuff of science fiction. I've included an edited excerpt of our conversation below, with slightly rephrased questions for a better flow.

Wray: What does extinction mean to you?

Van Dooren: That's actually a really big and difficult question. There's a tendency to mystify extinction, to think about it as the last individual of a kind. There are some really iconic examples, like Martha, the last passenger pigeon, who died at the Cincinnati Zoo, or the last thylacine in a Tasmanian zoo. There's this idea that the last individuals finally die and then we've seen the event of extinction, that their passing is the death of a species. On some level that's accurate, but on another level, and long before that last death, we see that all the relationships that animal was involved with, whether ecological relationships or human relationships with communities they may have provided meaning for, have broken down. So this is what I've been thinking about: the dullage of extinction. It is a long, slow unraveling of ways of life that happens long before the death of the last individual.

Wray: So extinction is a durational process rather than an instantaneous one?

Van Dooren: That's right. I think the focus on the individual does a lot of things and some are relevant to de-extinction. The focus on the individual in de-extinction, bringing the first one back, gives a sense that there's some ongoing continuity about the species—that it still survives in some way—even though it's not really the case. This was really drawn out for me by a visit I had to Hawaii, where I visited a snail ark. Tree snails are incredibly endangered. Where I visited, they're keeping the last of a whole bunch of snail species in a captive environment. Amidst them was this one little snail, the last of its kind. I sat and looked at it for quite a while, which was very depressing. But the idea that this species is ongoing in some way because this single individual exists in captivity—there's something misleading about thinking that its extinction hasn't yet taken place when its whole ecological role has disappeared.

Wray: Do we need to feel emotional about extinctions in order to learn from them?

Van Dooren: Personally I think it's definitely called for and an important part of relating to extinction. I don't think it's purely something that can be adequately fleshed out in cold, detached terms. I think extinctions call on us ethically for response, and those calls ought to be felt as well as rationalized. So what is significant about the disappearance of a species? For example, people who experience the disappearance of a species. I've worked on the Indian vulture, looking at the role they played in a small Parsi community in Mumbai. [The Parsi] dispose of their dead in Towers of Silence, and the vultures would eat them. But as the vultures have disappeared, the funerary practices have had to be revised, and according to many people in the communities, funerals are no longer functional.

I've talked to people who have gone through incredible suffering, who have left their dead to rot in Towers of Silence where the vultures should have come but didn't. These are incredibly emotional situations, and I think we can't fully engage with the significance of extinction if we try to give a cold and detached calculus of it. It's now thought that the absence of vultures in the environment may lead to an increase in street dogs, which will lead to increases in rabies. Part of mourning is coming to recognize those entanglements and how they matter.

Wray: How do you define mourning?

Van Dooren: I guess there's a lot of ways to answer that. On one level mourning is an evolved capacity that we most likely share with a whole range of intelligent social animals and consequently have become emotionally entangled in one another's lives. It is when a loved one disappears and we feel that loss. As Colin Murray Parkes famously put it, grief is "the cost of commitment." So to be at stake in others' lives emotionally exposes us to mourning when those relationships disappear. On an important level, mourning is part of our inheritance, our biogenetic inheritance, something we are lucky enough to share with other animals on the planet and a capacity we are failing to exercise. In the face of mass extinction, I think we are called on to learn how we are at stake in species of all sorts and to learn to become attuned to how these species matter and constitute the world for us. That should ultimately call for an emotional response—for grief and mourning, I would say.

I'm particularly interested in the capacity for mourning to change the world. There's a philosopher and grief counselor, Thomas Attig, who talks about mourning as a process of relearning the world. For him, it's a confrontation with a changed world. Through mourning we come to understand that something about the world we inhabited is gone, and that if we are going to go on, we must ourselves change; we must relearn the world. So mourning requires that we must reconsider how we fit in, and change might be demanded of us to go on. In the absence of mourning, we miss those opportunities for deep reflective work as individuals and as a culture about how we might go on differently.

Mourning introduces us to a relational conception of the world. It reminds us of all the ways we are connected to others for nourishment, meaning, cultural practices... all of the ways we are bound up in others. When we are mindful of those relationships, then the event of extinction or event of death unmakes us in some ways. It reminds us how we are vulnerable and have a stake in others. It calls for an awareness of our ecological embeddedness in broader environments. So responding in that way is about making sense of how we are relationally constituted and undone by the death of a species.

Wray: One could argue that if we caused the extinction of a species, and have the technology to recreate it, that we then have a moral duty to do so. What do you think about that?

Van Dooren: The idea that we have an obligation to bring them back because we caused their extinction is something that ought to be responded to because it is far too simple an idea. I think we'll continue to debate for decades about this question of duty, but at the end of the day ethics isn't just about identifying duties—it is also about how we inhabit a world of always conflicting duties. We need to think about case studies and the pros and cons of bringing back this extinct species. For example, no one would say we should bring smallpox back to the environment because we caused it to go extinct. So we will have to think about how a potential duty to bring back one species lines up with duties to humans and other animals and ecosystems wrapped up with that. It is too simple to say there's a duty because we caused them to go extinct, and I do also think there's a conflicting duty to respect the dead and let them rest in peace.

I've done some thinking about keeping faith with death. What that opens up for us is whether resurrection constitutes a restitution for some past harm done or whether that would be a desecration, in some sense, of the dead. That's especially the case in the world of ongoing loss and mass extinction. So bringing them back to violence and another extinction event is not, I think, the ethical thing to do. I think we are not yet ready for de-extinction, if we ever will be. I think we've shown ourselves to lack that cultural and political capacity to deal ethically with endangered species, and bringing more back, to me, is not the ethical response. The duty should be, at least for now, to leave these species to rest and instead to learn something from their extinction. I think we owe the dead to genuinely learn from them and change our ways so we don't send more over the edge into extinction.

Wray: What exactly do you mean by saying that we can keep "faith with death"?

Van Dooren: Keeping faith with an extinct animal is about not resurrecting them; it's about making the difficult choice to learn from their extinction, to mourn . . . and as a result change the way that we live. It is the attempt to not rush to overcome death and instead prevent it from happening again. Given the current context, as we're letting endangered species go extinct, resurrecting them doesn't represent the beginning of a new ethical relationship with them. It could just begin another phase of extinction for them.

Wray: What does it mean to have an ethics of extinction?

Van Dooren: De-extinction undermines our moral and imaginative capacity to engage with the current extinction crisis by allowing us the illusion that there is some techno-fix to our current situation and sidestep the vitally necessary cultural and political work that is needed to deal with this current extinction event. It spreads the idea that we can deal with it another day, that we don't need to worry urgently about the disappearance of every species—that we can bank them and set the world right at a later date. The capacity for that kind of technology to be captured by economic and political interests is, I think, a real threat to ongoing efforts to deal with conservation issues and the current mass extinction event in a concrete way.

All those anthropogenic causes are driven by economic and cultural incentives that need to be challenged and changed. It all calls for deep reflection on our values, ways of life, and priorities of all types. If we can avoid doing that kind of work and engineer our way around the solution with a techno-fix, we are likely to take that easy option. The idea that we shouldn't mourn but should do something is really worrying. Technical intervention is offered as an alternative instead of mourning. I think it gives us too much of an easy way out.

IN ONE OF his TED talks, Stewart Brand speaks about the tragedy we perceive when a species vanishes, and offers some advice: "Don't mourn, organize!" he bellows with a smile. When I later ask him what he meant by that, he says, "All you can do with extinction is just grieve. The animal's gone—it's *gone*. But if the technology is at a particular point where that kind of species can be brought back, then you want to bear down and make it happen."

Indeed, that pragmatic approach has become the name of the game in de-extinction, which values action over mourning or dwelling on our own survivor's guilt. Long before *de-extinction* was even a term, Brand subscribed to a particular form of pragmatism—eco-pragmatism—for which he laid out his own vision in his 2009 book *Whole Earth Discipline: An Ecopragmatist Manifesto.* Eco-pragmatism embraces technological interventions as solutions to environmental problems. From genetic engineering to geoengineering, an eco-pragmatist sees the benefits of technology for nature and does not shy away from applying it just because potential risks exist. But critics have said that eco-pragmatism is more interested in humanity's own abilities than anything else and that it will only march us dangerously toward an imaginary techno-utopia.

Environmental ethicist Ben Minteer identifies an acute problem with the eco-pragmatist approach as it applies to de-extinction—mainly that it flies in the face of old cultural values about ecosystem integrity. Historically, the integrity of an ecosystem depended on the interactions of species within the ranges those species were adapted to. The less touched by humans those areas were, the more pristine an ecosystem was. The dilemma here, according to Minteer, is that the era in which we now live—of increasing environmental intervention—creates thorny issues we must face when we contemplate our moral responsibility to species. The increasing artificiality, control, and manipulation of the landscape we engage in leads us to intensify the relocation of species in habitats outside of the areas they evolved in. It is in this assisted species colonization of new areas outside their home range where things start to get dicey, including the introduction of unextinct species. In this sense, an eco-pragmatist approach that promotes de-extinction and a rewilded world of translocated species unravels the values that once honored the integrity of ecosystems.

In a talk entitled "Extinction and the Price of Pragmatism," which he gave at the 2014 Forum on Ethics and Nature at the Chicago Botanic Gardens, Minteer opened with a quotation he pulled from a piece of hundred-year-old writing penned by William T. Hornaday—the first director of the Bronx Zoo. Hornaday was a crusader for wildlife protection in the late nineteenth and early twentieth centuries. In 1914 he wrote that "nature was a million years, or more, in developing the picturesque moose, the odd mountain goat, and the unique antelope. Shall we destroy and exterminate those species in one brief century? The young Americans of the year 2014 will read of those wonderful creatures, and if they find none of them alive how will they characterize the men of 1914? I, for one, do not wish in 2014 to be classed with the swine of Mauritius that exterminated the dodo."

So here's Minteer, exactly one century later, in the very year that Hornaday was envisioning, talking to a group of people about how we have failed to learn from the past. At least Hornaday's premonitions about some of those species did not come to fruition, but 1914 did go on to become the year that the passenger pigeon became extinct. Hornaday's visions of the future have an eerie quality about them. Elsewhere he wrote, "Let no one think for a moment that any vanishing species can at any time be brought back; for that would be a grave error ... *The heath hen could not be brought back, neither could the passenger pigeon* [emphasis added]." Little did he know what would come of that far-fetched idea in just one hundred years.

Today, Minteer says, as we take a more action-oriented and progressive approach toward conserving species and perhaps even recreating them, we are forced to make tough trade-offs. Traditional preservationist values focus on maintaining the historical integrity of ecosystems and are challenged by ideas of translocation, reintroduction, and unextinct species, which make us increasingly dependent on novel landscapes that we have curated and arranged. And as this happens, our conservation ethics will increasingly hinge on a "nature by design." Minteer urges us to think carefully about that. And when I do, I do not discover a moral belief deep within me that says ecosystems should not change at the hands of humans; they do that all the time. After all, that's the essence of living in these times, although it's clear that we have to get a lot better at how we go about it.

Instead, thinking about this forces me to acknowledge that designed natural areas put an extra burden on us to understand exactly how we need to meticulously plan, organize, and manage those areas rather than let them evolve in a more haphazard way. This places demands on us to know how to best care for our "monsters" that we create through de-extinction, as well as the valued habitats we put them into. We're liable for whatever could go wrong, at the same time as we are aware that things could always be done otherwise, or not be done at all. That's partly why de-extinction is such a fascinating site for understanding how science and society collide. What moves are scientists personally willing to make? About what does society disagree? And how will we deal with the fallout?

From writing this book, I've learned that I want to develop a fuller account in my mind of what happens to life when a particular form of it disappears. When a singular species goes extinct, a whole way of being in the world vanishes along with it, and a flame of existence is forever extinguished. No amount of my mourning, misplaced or not, can erase that fact. Necrofauna will not rise from the dead—their proxies will be created anew—and extinction cannot be reversed. What we're left with instead, in every case of extinction and de-extinction, will be an ultimately different and changed world. If we skate over the meaning of extinction or what it means to undo extinction, we could miss out on the opportunity to realize our own entanglements with species and their unique condition for being in the world. Jennifer Jacquet's emotional distress about the sea cow's disappearance might be worth all those sleepless nights after all. It gives her time in the middle of the night to ponder how its extinction transforms the world.

AN AMERICAN ARTIST—and friend to Revive & Restore— named Isabella Kirkland paints endangered and extinct animals in large Dutch-master-style oil paintings. She begins each painting with a catalog that marks out which species she's working with are extinct, which are invasive, and which are newly discovered taxa. She uses real specimens to capture their likeness, and has held the bones of the laughing owl, the horns of the bluebuck, and even the eggs of the Syrian ostrich for her visual research. Her painting *Gone* has sixty-three animals in it, every one of which has disappeared since the colonization of the so-called New World. And as de-extinction advances, her painting serves as a time capsule to be viewed in a future when some of these animals may reappear in recreated forms. In *Gone*, a bright yellow and green Carolina parakeet—a species that died out in 1918—is perched on a twig. Below it sits the tiny golden toad, the last of which was seen in the Costa Rican rainforests in 1989. Front and center up high sits the most striking creature: Martha—the last passenger pigeon.

If Ben Novak eventually succeeds in his mission to recreate Martha's species, I wonder, will Kirkland go back to the painting and update it, taking Martha out? If so, she'd have to find some new animal to put in the center, though that shouldn't pose much of a problem. There are, after all—and will continue to be—lots of extinct species that could fit. Even though we may have discovered how to resurrect the traits of extinct creatures, we haven't figured out how to stop the human-caused ecological destruction that keeps sending more species over the edge of extinction. But if the passenger pigeon will still be extinct in Kirkland's eyes no matter the degree of Novak's success in making a band-tailed pigeon full of passenger pigeon genes, perhaps she wouldn't bother. Art is subjective, and it would seem that sometimes science can be too.

Kirkland painted *Gone* as a reminder of what's been lost, but the painting has been repurposed as part of de-extinction's public face. It welcomes every visitor to Revive & Restore's website, bold and beautiful, filling its landing page. At the same time, *Gone* also graces the cover of an extinction studies compendium coedited by Thomas van Dooren, who impressed upon me the importance of "keeping faith with the dead" as the basis for opposing de-extinction. The painting means different things to different people, just like de-extinction itself.

And when people see *Gone,* Kirkland wants the artwork to make them stop and think, "What really matters here?" When she first heard about de-extinction, she tells me, she thought, "Well, that's quixotic! It's a very odd thing to consider… (a) It's impossible, (b) It will cost too much money, and (c) You know, people are going to get upset about GMOs and whatnot." But the more she considered it, the more excited she became. What initially seemed impossible ripened into hopes for something real. Reflecting on it later, she says, "We are animals that rely on a certain amount of hope, and I find this whole idea of de-extinction very, very hopeful."

I agree that we are animals that rely on a certain amount of hope. But if Kirkland were ever to repaint this menagerie of extinct species as it lives inside my mind, I do not think that my own sense of hope would be restored by taking Martha out of it. I firmly believe that we need to remember and honor the differences between species that we have made disappear in the past and species that we are now forcing into existence in the name of ecological justice, without their even knowing it.

Today, the world I inhabit after spending so much time with de-extinction is populated with creatures of a curious kind, and if feelings were organisms, ambivalence would be the keystone species I interact with most often. What I've learned from the species I've encountered during this journey—as the ones we've known disappear and new ones are prepared to come to life— is that the more we try to let them off their leashes, the more tightly tethered to them we actually become. There's a deep irony here: the wild animals we're creating won't be able to make it in the wild without our help. For if these creatures are ever to flourish in the great outdoors as they're imagined to, we must plan, design, monitor, manage, and maintain their ways of life. This creates novel conditions for their wildness that the extinct species they're meant to mimic never needed.

Learning about de-extinction has sent me into the heart of an ecosystem that's colonized with even more mixed feelings than there are mixed genes. And while some say that de-extinction is being developed for the well-being of ecosystems we've impoverished—or ones about to slip away—there's only one species in any of them that gets to ask the most interesting question: Who gets to live and who gets to die in the environments that de-extinction is devising?

So, how should we think about the future of this human-mediated nature? As Dolly Jørgensen taught me, the narratives that we use to do this thinking matter. The idea that we might see fauna from the Pleistocene walk the Earth again lays the ground for a fantastic, mind-boggling, and awe-inspiring story. I'm sure that's why two production companies have expressed interest in transforming this book into a film, and other people's species resurrection films are on the way. The mere possibility of de-extinction is entertaining, and it effortlessly captures our hearts and minds. But the thing is, framing the story of de-extinction from any one particular perspective masks other stories that could have been told instead. What about the stories from the species themselves? I wonder what the passenger pigeon, woolly mammoth, gastric-brooding frog, thylacine, and aurochs would say about all of this if they could talk. Would they feel genuinely loved, missed, and cared for by our efforts to de-extinct them? Or would they think that our motivations are somehow perversely misplaced? When we tell stories about de-extinction, should we speak of salvation, resurrection, and atonement for our past sins? Should we talk about environmental justice and ecological repair? Is it better to focus on love and mourning, to learn how to care better for the future? Or centralize the narrative on a straightforward quest to push the technology and research further?

Whatever we choose, we need to step up to the plate of responsibility and reevaluate the possibilities for what comes next. De-extinction will give birth to much more than unextinct creatures in a lab. It will give birth to new social and scientific practices, cultures, laws, attitudes, values, and beliefs. We must seriously ask ourselves what de-extinction will do to species, but we should not forget to also ask, what will de-extinction do to *us*? What does it teach us about who we are and how we want to live in this world?

At the beginning of my research, I was a bit skeptical about de-extinction and what it might do. I felt that there was a lot of hubris surrounding the mere suggestion that we should be recreating extinct species when there are so many unknowns involved and creatures' lives at stake. I presumed the movement would play favorites about which species to revive and, in doing so, would risk being more about our own legacy than that of the species we have made disappear. I was unconvinced that the promises de-extinction was making were anything more than wishful techno-fixes, and I was curious about the possibility that unspoken motivations may be behind some of it.

However, as I spent more and more time with the people who have spent years developing de-extinction—and remember, although a handful of very talented people are actively doing this, it's a relatively fringe scientific activity—I realized how earnestly they believe that de-extinction will make a beneficial difference in the world, have complete faith that it will help the ecosystems and species that they adore, and swear by its ability to restore people's sense of hope for life on this planet in times of environmental decay. I am now a lot more trusting of their motives and admiring of their efforts. But I see de-extinction—understood as the re-creation of extinct species—as a fascinating technological feat more than an environmental necessity. Genetic engineering will become a vital tool for conservation in many important respects, but I don't regard the re-creation of extinct species as a priority among them. In this sense I'm more excited about the technology's application to endangered species than I am about its use for those already gone. But as we've seen with cases like the Woolly Mammoth Revival project, de-extinction may serve dual purposes depending on how it is applied: resurrecting extinct mammoth genes while simultaneously boosting populations of endangered elephants by giving them beneficial traits. What remains to be understood is how much environmental benefit this will create in relation to how much it will cost (in all senses of the word).

Some have said that de-extinction will happen whether we like it or not, and have written things such as "inevitable technological development means that extinct species will be resurrected at some future point." But I would like to point out that de-extinction is still just an option, as there will always be a human at the helm who decides whether it will go forward or not. I think our biggest challenge, if we are to pursue it fully and with increasing fervor, is to somehow couple de-extinction with improved strategies to overcome the larger structural issues that endanger species in the wild. Without expanding on the hard work that conservationists, environmentalists, and some politicians have been doing for decades, de-extinction risks being done in vain. We must be alert to how de-extinction is developing in practice and expect that things will not always go as planned. But if we choose to use it, it is our responsibility to ensure that it will create a truly better world and not merely a more ecologically complex one.

ACKNOWLEDGMENTS

I'D LIKE TO start by saying what an unpredictable and enjoyable adventure this entire exploration of de-extinction has been. It all began on a dance floor on the outskirts of Chicago at the Third Coast Audio Festival, where I met Sara Robberson Lentz, who would later become my collaborator on a radio production about "the art of de-extinction" that we made for the Science and Creativity series on WNYC's *Studio 360*. Thank you, Sara, for taking that first plunge with me, and thank you, Ann Heppermann, for being a boon to the piece with your edits.

The next nesting ground for my work on the subject was also in radio, but this time at the "Mothercorp," as we Canadians call it: the CBC. When the team at CBC Radio One's *Ideas* accepted my pitch for an hour-long feature documentary about de-extinction, I had no idea what I was truly gaining from the opportunity. Not only was it an honor to produce for their national show, but because my story attracted the ears of a particularly skilled staff producer with a passion for science, I ended up many times more fortunate from the commission: I met my CBC radio mentor, Sara Wolch, who I am so grateful to have been trained by. Thank you, Sara, for not going easy on me in the editing room. Without these radio productions, there would be no book.

And neither radio production would have been possible without the many experts and inspiring individuals I was able to call on for their ideas, knowledge, and opinions. I must, therefore, extend a big thank-you to the owners of the voices that weave those pieces into what they are: Stewart Brand, Ben Novak, Isabella Kirkland, David Ehrenfeld, Beth Shapiro, Hendrik Poinar, Dolly Jørgensen, Norman Carlin, Thomas van Dooren, and Mark Peck.

Then a huge surprise occurred one day when I got a one-liner in my email that read something to the tune of "Great radio show, now how about a book?" signed by the head of a publishing company. I had to triple check to make sure that this wasn't a piece of auto-drafted spam before I took it seriously. And am I ever delighted that it wasn't. Thank you a thousand times to Rob Sanders, publisher at Greystone Books, for taking a leap of faith in me and sensing that I would be up for the assignment. Your great enthusiasm for the subject has been hugely supportive from the start. My very talented editor at Greystone Books, Nancy Flight, has been a joy to work with, and these pages have benefited in uncountable ways by her sharp sense that always sees how a sentence can still be clearer. I am so grateful for the efforts that the entire team at Greystone Books—from marketing and communications to design and administration—has poured into this project. And I am humbled that my book is being published in cooperation with the David Suzuki Institute as a result of working with this very special publisher.

It is a great honor that George Church, one of the world's most revered scientists working not only in de-extinction but in genetics as a whole, agreed to write such a stimulating foreword for this book—thank you very much, George.

I would not know much about de-extinction at all if it weren't for the many incredibly smart sources I met as I was researching this book and my other productions. I owe much gratitude to everyone who agreed to speak with me and/or write to me about the ideas explored in these pages, sometimes on several occasions, sometimes taking a lot of their time. These people (in no particular order) are Ryan Phelan, Stewart Brand, George Church, Ben Novak, Hendrik Poinar, Bobby Dhadwar, Dolly Jørgensen, Thomas van Dooren, Andrew Torrance, George Poinar Jr., Henri Kerkdijk-Otten, Nikita Zimov, Tom Gilbert, Stuart Pimm, Phil Seddon, Vincent Lynch, Josh Donlan, Beth Shapiro, James Mwemba, Joel Sartore, Luke Griswold-Tergis, Norman Carlin, Mark Peck, Max Holmes, Matt Ridley, Donna Haraway, Kent Redford, David Ehrenfeld, Joseph Bennett, Isabella Kirkland, Marguerite Humeau, Persephone Pearl, Carl Zimmer, Mikkel Holger Strander Sindig, Heath Packard, Zack Denfeld, Cat Kramer, Maja Horst, and Sarah Davies.

Thank you also to the lovely people at Quercus Group, who housed us while we were in Kenya meeting rhinos.

A big thank-you goes out to Charles Yao, for sharing his excellent insights about the world of publishing, and, by extension, the advice of Kelvin Kong. I owe the same to the wonderful Eliza Clark and Laurel Braitman. I sincerely appreciate the time my friend Anders Meldgaard Kjemtrup put in to critique the first draft, as well as Tom Gilbert and Johan Andersen-Ranberg, who lent their scientific expertise as reviewers of the entire manuscript. A huge thank-you also goes out to Ben Novak, who kindly reviewed the chapter that he most prominently appears in, and to Chris Widga, who generously reviewed the chapter about mammoths. I am bowled over by the attention span of my fact-checker Chris Morgan. Thank you also to Stephanie Fysh and Dawn Loewen, whom I had the pleasure of working with at the copy editing and proofreading stages respectively.

Thank you to my amazing dad, Joe Wray, whose long talks with me on the weekends during my childhood about science and ethics whetted my appetite for doing the work that I do now, and thank you to Nancy Jane Wray for being the whip-smart life-long conversant with us on these topics that she is. I endlessly admire the ability of my mom, Toni Wray, to support and be resilient, which helped me get through this project as I crazily took it on right when I was starting a PhD about a different subject. My deepest gratitude goes to my love, Sebastian Damm, the most charismatic creature of them all.

NOTES

EACH NOTE HERE refers to a direct in-text quotation that comes from another source. Where quotations are not listed here, they came from interviews that the author personally conducted.

INTRODUCTION

"To Bring Back the Extinct" "To Bring Back the Extinct: A Conversation with Ryan Phelan," Edge, accessed August 29, 2012, https://www.edge.org/conversation/ryan_phelan-to-bring-back-the-extinct.

"One of the fundamental" Ibid.

"The big question that" Ibid.

"We're using the term" Ibid.

"*resurrection ecology*" Brian Switek, "The Promise and Pitfalls of Resurrection Ecology," *National Geographic Phenomena,* March 13, 2013, http://phenomena.nationalgeographic.com/2013/03/12/the-promise-and-pitfalls-of-resurrection-ecology/.

"*species revivalism*" Jamie Shreeve, "Species Revival: Should We Bring Back Extinct Animals?" *National Geographic News,* March 6, 2013, http://news.nationalgeographic.com/news/2013/03/130305-science-animals-extinct-species-revival-deextinction-debate-tedx/.

"*zombie zoology*" Sandra Swart, "Frankenzebra: Dangerous Knowledge and the Narrative Construction of Monsters," *Journal of Literary Studies* 30, no. 4 (October 2014): 45-70, doi:10.1080/02564718.2014.976456.

"charismatic necrofauna" Jason Mark, "Back from the Dead," *Earth Island Journal*, Autumn 2013, http://www.earthisland.org/journal/index.php/eij/article/back_ from_the_dead/.

"it was sort of like" Steve Jobs, Commencement Address, Stanford University, *Stanford News*, June 14, 2005, http://news.stanford.edu/2005/06/14/jobs-061505/.

"We are as gods and might as well" Stewart Brand, "Purpose," *Whole Earth Catalog*, Fall 1968, accessed May 2, 2017, https://monoskop.org/images/0/09/Brand_Stewart_ Whole_Earth_Catalog_Fall_1968.pdf.

"We are as gods and have to" "We Are as Gods and Have to Get Good at It: Stewart Brand Talks about His Ecopragmatist Manifesto," Edge, accessed January 18, 2015, http:// edge.org/conversation/we-are-as-gods-and-have-to-get-good-at-it.

"De-extinction is essentially a game changer" Stanley Temple in Stewart Brand, meeting report, Forward to the Past: De-extinction Projects, Techniques, and Ethics, Washington, DC, October 23–24, 2012, Revive & Restore, accessed January 9, 2016, http://longnow.org/revive/1stde-extinction/.

"The pigeon was a biological storm" Aldo Leopold, "On a Monument to the Pigeon," *A Sand County Almanac* (Oxford: Oxford University Press, 1949), 108–12.

"Dear Ed and George" Nathaniel Rich, "The Mammoth Cometh," *New York Times Magazine*, February 27, 2014, http://www.nytimes.com/2014/03/02/magazine/ the-mammoth-cometh.html.

"a flock of millions to billions" Ibid.

"to enhance biodiversity" "What We Do," Revive & Restore, accessed March 18, 2015, http://longnow.org/revive/what-we-do/.

"the trick will be" "What Is Genetic Rescue?" Revive & Restore, accessed March 19, 2015, http://longnow.org/revive/what-we-do/genetic-rescue/.

"The precept against taking life" Gary Snyder, *The Practice of the Wild* (Berkeley, CA: Counterpoint Press, 2010), p. 195.

"Humans have made a huge hole" Stewart Brand, "The Dawn of De-extinction. Are You Ready?" (video), TED, February 2013, https://www.ted.com/talks/stewart_brand_ the_dawn_of_de_extinction_are_you_ready.

"I probably should have called" Beth Shapiro in Ewen Callaway, "Mammoth Genomes Provide Recipe for Creating Arctic Elephants," *Nature* 521, no. 7550 (May 1, 2015): 18, doi:10.1038/nature.2015.17462.

"I think there's a huge misconception" Beth Shapiro in Diane Toomey, "Cloning a Mammoth: Science Fiction or Conservation Tool?" Environment 360, Yale University, June 17, 2015, http://e360.yale.edu/feature/cloning_a_mammoth_ science_fiction_or_conservation_tool/2886/.

CHAPTER ONE: HOW IS DE-EXTINCTION DONE?

"You can't clone from stone" Robert Lanza, "The Use of Cloning and Stem Cells to Resurrect Life" (video), TEDxDeExtinction, April 2, 2013, https://www.youtube.com/watch?v=gqpErbFnbiY.

"If society becomes comfortable" George M. Church and Ed Regis, *Regenesis: How Synthetic Biology Will Reinvent Nature and Ourselves* (New York: Basic Books, 2012), p. 11.

"It had been crudely stuffed" Reinhold Rau, "Rough Road towards Re-breeding the Quagga: How It Came About," March 1999, accessed November 29, 2016, http://www.quaggaproject.org/downloads/2012/rough%20road%20to%20re%20breeding%20quagga-1.pdf.

"Unfortunately, as I suspected" Ibid.

"struck like a bombshell" George Poinar Jr., "Ancient DNA: The Tools of Molecular Biology Can Be Used to Peer into an Organism's Genetic Future—or Its Distant Past," *American Scientist* 87, no. 5 (September 1, 1999): 446–57, http://www.jstor.org/stable/27857905.

"Nazi super-cows" Steven Morris, "Devon Farmer Forced to Offload Aggressive Nazi-Bred 'Super Cows,'" *The Guardian*, January 5, 2015, https://www.theguardian.com/world/2015/jan/05/devon-farmer-forced-offload-nazi-bred-super-cows.

"Rau quaggas" The Quagga Project, accessed January 3, 2017, http://quaggaproject.org/.

"Today we go barbecue... mammoth" *Woolly Mammoth: The Autopsy* (television program), written and directed by Nick Clarke Powell, Renegade Pictures, broadcast on Channel 4, November 23, 2014, accessed March 20, 2015, http://www.channel4.com/programmes/woolly-mammoth-the-autopsy.

"mugshots" Eugene Koonin in "Antibodies Part 1: CRISPR" (podcast), *Radiolab*, accessed January 16, 2016, http://www.radiolab.org/story/antibodies-part-1-crispr/.

"In parallel with the arrival" Stewart Brand, "Rethinking Extinction," *Aeon*, April 21, 2015, https://aeon.co/essays/we-are-not-edging-up-to-a-mass-extinction.

"Not since J. Robert Oppenheimer" Michael Specter, "The Gene Hackers," *The New Yorker*, November 16, 2015, http://www.newyorker.com/magazine/2015/11/16/the-gene-hackers.

CHAPTER TWO: WHY IS DE-EXTINCTION IMPORTANT?

"The observer would see" Manuel DeLanda in Erik Davis, "DeLanda Destratified," *Mondo 2000*, no. 8 (Winter 1992), accessed January 6, 2017, https://techgnosis.com/de-landa-destratified/.

"any substantial increase in" J.J. Sepkoski Jr., "Phanerozoic Overview of Mass Extinction," in *Patterns and Processes in the History of Life*, edited by D. M. Raup and D. Jablonski, Dahlem Workshop Reports 36 (Berlin; Heidelberg: Springer, 1986), p. 278.

"without any significant doubt" Gerardo Ceballos, Paul R. Ehrlich, Anthony D. Barnosky, Andrés García, Robert M. Pringle, and Todd M. Palmer, "Accelerated Modern Human-Induced Species Losses: Entering the Sixth Mass Extinction," *Science Advances* 1, no. 5 (June 1, 2015), doi:10.1126/sciadv.1400253.

"Speciation by hybridization" Chris D. Thomas, "The Anthropocene Could Raise Biological Diversity," *Nature* 502, no. 7469 (October 2, 2013): 7, doi:10.1038/502007a.

"Any creature or plant facing" Brand, "Rethinking Extinction."

"Viewing every conservation issue" Ibid.

"*hyperobjects*" Timothy Morton, *Hyperobjects: Philosophy and Ecology after the End of the World* (Minneapolis: University of Minnesota Press, 2013).

"The core of tragedy is" Brand, "Rethinking Extinction."

"not just money" International Union for the Conservation of Nature (IUCN), *IUCN SSC Guiding Principles on Creating Proxies of Extinct Species for Conservation Benefit*, May 18, 2016, https://portals.iucn.org/library/sites/library/files/documents/Rep-2016-009.pdf.

CHAPTER THREE: WHAT SPECIES ARE GOOD CANDIDATES FOR DE-EXTINCTION, AND WHY?

"Maybe we can edit" Stewart Brand, "De-extinction Debate: Should We Bring Back the Woolly Mammoth?" Point/Counterpoint, Yale Environment 360, January 13, 2014, http://e360.yale.edu/feature/the_case_for_de-extinction_why_we_should_bring_back_the_woolly_mammoth/2721/.

"fascinating but dumb idea" Paul R. Ehrlich, "The Case against De-extinction: It's a Fascinating but Dumb Idea," Point/Counterpoint, Yale Environment 360, January 13, 2014, http://e360.yale.edu/feature/the_case_against_de-extinction_its_a_fascinating_but_dumb_idea/2726/.

"Resurrecting a population" Ibid.

"a giant, weird duck" Michael Archer, "How We'll Resurrect the Gastric Brooding Frog, the Tasmanian Tiger" (video), TEDx Talks, March 2013, https://www.ted.com/talks/michael_archer_how_we_ll_resurrect_the_gastric_brooding_frog_the_tasmanian_tiger.

"great big ones to middle-sized" Ibid.

"The length of this animal" Robert Paddle, *The Last Tasmanian Tiger: The History and Extinction of the Thylacine* (Cambridge: Cambridge University Press, 2002), p. 4.

"Study of ancient DNA was" Svante Pääbo, "Preservation of DNA in Ancient Egyptian Mummies," *Journal of Archaeological Science* 12, no. 6 (November 1, 1985): 411–17, doi:10.1016/0305-4403(85)90002-0.

"When it launched the thylacine" Amy Fletcher, "Genuine Fakes: Cloning Extinct Species as Science and Spectacle," *Politics and the Life Sciences* 29, no. 1 (March 1, 2010): 31, doi:10.2990/29_1_48.

"At the outset" Ibid., p. 54.

"Love the idea or hate it" Philip J. Seddon, "De-extinction: Reframing the Possible," *Trends in Ecology & Evolution* 30, no. 10 (October 2015): 570. doi:10.1016/j. tree.2015.08.002.

"1. Can the past cause(s)" Philip J. Seddon, Axel Moehrenschlager, and John Ewen, "Reintroducing Resurrected Species: Selecting DeExtinction Candidates," *Trends in Ecology & Evolution* 29, no. 3 (March 2014): 140-47, doi:10.1016/j.tree.2014. 01.007.

CHAPTER FOUR: WHY RECREATE THE WOOLLY MAMMOTH?

"The work will attract funding" Kent H. Redford, William Adams, Rob Carlson, Georgina M. Mace, and Bertina Ceccarelli, "Synthetic Biology and the Conservation of Biodiversity," *Oryx* 48, no. 3 (July 2014): 333, doi:10.1017/S0030605314000040.

"It would be awe inspiring" Hank Greely, "De-extinction: Hubris or Hope?" (video), TEDx-DeExtinction, April 1, 2013, https://www.youtube.com/watch?v=HuRkoV2LoMY.

"And that is a real advantage" Ibid.

"For all my protests" Tori Herridge, "Mammoths Are a Huge Part of My Life. But Cloning Them Is Wrong," *The Guardian*, November 18, 2014, http://www.theguardian.com/ commentisfree/2014/nov/18/mammoth-cloning-wrong-save-endangered-elephants.

"A few dozen changes to the genome" George Church, "George Church: De-extinction Is a Good Idea," *Scientific American*, September 1, 2013, http://www.scientificamerican. com/article/george-church-de-extinction-is-a-good-idea/.

"When your dog has passed away" Sooam Biotech Research Foundation, accessed November 14, 2015, http://en.sooam.com/.

"In other words, we cannot" Helen Pilcher, "Reviving Woolly Mammoths Will Take More than Two Years," BBC, February 22, 2017, accessed February 28, 2017, http://www. bbc.com/earth/story/20170221-reviving-woolly-mammoths-will-take-more-than-two-years.

"This generation gets to rethink" Stewart Brand, "Bringing Back the Passenger Pigeon," Revive & Restore, February 22, 2012, http://reviverestore.org/passenger-pigeon-workshop/.

CHAPTER FIVE: CAN BILLIONS OF PASSENGER PIGEONS REBOUND, AND SHOULD THEY?

"mourning doves on steroids" Joel Greenberg, *A Feathered River across the Sky: The Passenger Pigeon's Flight to Extinction* (New York: Bloomsbury, 2014), p. 1.

"To make a pot pie of them" Quoted in Margaret Mitchell, *The Passenger Pigeon in Ontario* (Toronto: University of Toronto Press, 1935), p. 21.

"These outlaws to all moral sense" Simon Pokagon, "The Wild Pigeon of North America," *Chautauquan* 22 (1895): 205.

"of the countless thousands of birds" Etta S. Wilson, "Personal Recollections of the Passenger Pigeon," *The Auk* 51, no. 2 (1934): 166, doi:10.2307/4077888.

"In packing for shipment" Quoted in Arlie William Schorger, *The Passenger Pigeon, Its Natural History and Extinction* (Madison: University of Wisconsin Press, 1955), p. 147.

"The reality of the passenger pigeon" Ben Novak in "De-extinction: What Does It Mean?" Google Hangout (video), September 24, 2014, "Empty Skies: The Passenger Pigeon Legacy," Royal Ontario Museum, accessed January 11, 2017, http://www.rom.on.ca/en/exhibitions-galleries/exhibitions/past-exhibitions/passenger-pigeons.

"People get scared of the idea" Ibid.

"Passenger pigeons and fire were the major sources" "The Great Passenger Pigeon Comeback," Revive & Restore, June 9, 2015, http://longnow.org/revive/projects/the-great-passenger-pigeon-comeback/.

"A new passenger pigeon may cause" Ben Novak, "Reflections on Martha's Centennial," Revive & Restore, September 1, 2014, http://reviverestore.org/of-martha-the-last-passenger-pigeon/.

"If you come for their Feathers" Quoted in Errol Fuller, *The Great Auk* (New York: Abrams, 1999), p. 401.

"The razorbill has a similar range" Stewart Brand, "2015 Year End Report," Revive & Restore, December 29, 2015, http://reviverestore.org/2015-year-end-report/.

"Two years ago at the 2014 Avian Model" Ben J. Novak, "The CRISPR Craze Takes Fight: Adding Birds to the CRISPR Zoo," Revive & Restore, April 14, 2016, http://reviverestore.org/the-crispr-craze-takes-fight-adding-birds-to-the-crispr-zoo/.

CHAPTER SIX: HOW MIGHT WE REGULATE THIS NEW WILDERNESS?

This chapter is fully composed of research that I collected from personally conducted interviews.

CHAPTER SEVEN: CAN DE-EXTINCTION SAVE SPECIES ON THE BRINK?

"According to Bat Conservation International" "Bats Are Important," Bat Conservation International, accessed November 2016, http://www.batcon.org/why-bats/bats-are/bats-are-important.

"Each cell of an individual is capable" Oliver Ryder in Al Jazeera America, "The Frozen Zoo," *TechKnow*, 2014, accessed November 19, 2015, https://www.youtube.com/watch?v=39RUr0s-4DQ.

"Is rescuing a species or a subspecies" Joseph Saragusty, Sebastian Diecke, Micha Drukker, Barbara Durrant, Inbar Friedrich Ben-Nun, Cesare Galli, Frank Göritz, et al., "Rewinding the Process of Mammalian Extinction," *Zoo Biology* 35, no. 4 (May 1, 2016): 8, doi:10.1002/zoo.21284.

"allowing that mosquito parent" Jad Abumrad in "Update: CRISPR" (podcast), *Radiolab*, accessed March 27, 2017, http://www.radiolab.org/story/update-crispr/.

"a gene that could spread through" Heidi Ledford and Ewen Callaway, "'Gene Drive' Mosquitoes Engineered to Fight Malaria," *Nature News*, November 23, 2015, doi:10.1038/nature.2015.18858.

CHAPTER EIGHT: IS SOME KNOWLEDGE TOO DANGEROUS?

"I am the resurrection, and the life" John 11:25, *The Holy Bible, King James Version*, Cambridge Edition (1769), *King James Bible Online*, www.kingjamesbibleonline.org.

"If we destroyed part of Eden" "The Scientist Trying to Reverse Extinctions (Think 'Jurassic Park')," *FiveThirtyEight*, November 25, 2015, http://fivethirtyeight.com/features/the-scientist-trying-to-reverse-extinctions-think-jurassic-park/.

"I'm Mike Archer" Ibid.

"Creation stories matter" Swart, "Frankenzebra," p. 45.

"Frankenstein complex" Isaac Asimov, "The Machine and the Robot," in *Science Fiction: Contemporary Mythology, The SFWA-SFRA Anthology*, edited by P.S. Warrick, M.H. Greenberg, and J.D. Olander (New York: Harper and Row, 1978).

"yellow, watery, but speculative eyes" Ibid.

"within the crowded market" Australian Museum, annual report 1999/2000, in Fletcher, "Genuine Fakes," p. 55.

"neoliberal necromancy" Swart, "Frankenzebra," p. 58.

"mythic narrative strength may obscure" Ibid., p. 49.

"Remember that I am thy creature" Mary Shelley, *Frankenstein: or, the Modern Prometheus: The Original 1818 Text*, 2nd edition (Peterborough, ON: Broadview Press, 1999), chapter 10, p. 3.

"I was benevolent and good" Ibid.

"My talk is going to be quite literal" Jennifer Jacquet in "Extinction Marathon: Visions of the Future" (exhibition), Serpentine Sackler Gallery, accessed July 18, 2016, http://www.serpentinegalleries.org/exhibitions-events/extinction-marathon.

"I decided to challenge the group" Lee Silver in Gregory E. Kaebnick, *Humans in Nature: The World as We Find It and the World as We Create It* (New York: Oxford University Press, 2013), p. 12.

"The attention to this topic" Kaebnick, *Humans in Nature*, p. 23.

"A purely emotional approach" Ibid., p. 22.

"Don't mourn, organize!" Brand, "The Dawn of De-extinction."

"nature was a million years" William Temple Hornaday, *Wildlife Conservation in Theory and Practice: Lectures Delivered before the Forest School of Yale University, 1914* (New Haven, CT: Yale University Press, 1914), p. 166.

"Let no one think for a moment" William Temple Hornaday, *Our Vanishing Wild Life: Its Extermination and Preservation* (New York: Scribner's, 1913), p. 323.

"inevitable technological development" Seddon, Moehrenschlager, and Ewen, "Reintroducing Resurrected Species," p. 140.

GLOSSARY

Artificial selection: the guided and intentional reproduction of individuals in a population that have desirable traits, often achieved through breeding techniques.

Backbreeding: a type of artificial selection that aims to recreate the traits of a wild-type ancestor in new individuals in a population, where the wild-type ancestor is usually extinct.

Chromosome: an organized structure that contains tightly coiled DNA of an organism.

Clone: an organism that is a near genetic copy of another organism.

CRISPR/Cas9: A naturally occurring immune system in bacteria and archaea that is programmable as a gene-editing tool.

Cryopreservation: a process by which cells, tissues, and other biological materials that are vulnerable to damage are preserved at very low temperatures.

Cytoplasm: a thick solution inside of cells, enclosed by the cell membrane but outside of the nucleus.

De-extinction: the overarching movement or practices of creating facsimiles of extinct species.

Diploid: the state of having two complete sets of chromosomes, one from each parent.

DNA: a self-replicating molecule that carries the genetic information of an organism.

DNA sequencing: *see* Sequencing.

DNA synthesis: *see* Synthesis.

Epigenetics: the study of how heritable changes are made in an organism when they cannot be explained by changes in the genetic code itself, for example, through gene expression as it is modified by environmental factors.

266

Eukaryotic cells: Cells that contain a nucleus and other organelles that are bound by membranes.

Gamete: a haploid male or female germ cell (for example, sperm or egg) that is able to fuse with another of the opposite sex in sexual reproduction to create a new diploid organism.

Gene: a unit of heredity that is transferred from parents to offspring; it is made up of a specific DNA sequence.

Gene editing (may also refer to genome editing): a type of genetic engineering in which DNA is deleted, inserted, or replaced in the genome of an organism.

Genetic engineering: the human manipulation of the DNA of an organism using biotechnological tools.

Genome: the complete collection of genetic material in an organism.

Germ cells: cells that give rise to the gametes (sperm and eggs) of an organism that reproduces sexually.

Haploid: the state of having a single set of unpaired chromosomes.

Induced pluripotent stem cells (iPSCs): cells that are reprogrammable to become embryonic stem cell–like.

Mitochondrion: a membrane-bound organelle found in large numbers in most eukaryotic cells that acts as the "power house" for the cell; it contains a small genome independent from that of the nucleus.

Nucleus: a membrane-bound organelle in eukaryotic cells that contains most of the cell's genetic material.

Pluripotent: the immature state of a cell that is able to give rise to all different cell types in the body.

Polymerase chain reaction (PCR): a method of making multiple copies of a DNA sequence.

Primordial germ cells (PGCs): the predecessors of germ cells that will eventually become gametes.

Recombinant DNA: DNA that has been directly formed by combining genetic material from different organisms.

Selective breeding: a type of artificial selection in which humans purposefully breed chosen animals or plants to develop desirable traits in individuals in a population.

Sequencing (DNA sequencing): the process of deciphering the precise order of DNA bases in a genetic sequence.

Somatic cell nuclear transfer: a laboratory technique for creating an egg cell with a donor nucleus; used in cloning.

Somatic cells: any type of nonreproductive cells that form the body of an organism.

Stem cells: unspecialized cells that can differentiate into specialized cells to form specific body parts.

Synthesis (DNA synthesis): the physical creation of artificial gene sequences.

Totipotent: the immature state of a cell that is able to give rise to any different cell type in the body as well as placental cells. Embryonic cells within the first couple of divisions after fertilization are the only totipotent stem cells an organism will have.

Unextinct: the state of having been created by a de-extinction process.

SELECTED BIBLIOGRAPHY

Adams, William M. "Geographies of Conservation I: De-extinction and Precision Conservation." *Progress in Human Geography,* May 18, 2016. doi:10.1177/0309132516646641.

Agence France-Presse. "Chinese Police Alleged to Have Eaten Endangered Giant Salamander at Banquet." *The Guardian,* January 27, 2015. http://www.theguardian.com/world/2015/jan/27/chinese-police-eaten-endangered-giant-salamander-banquet.

Ajmone-Marsan, Paolo, José Fernando Garcia, and Johannes A. Lenstra. "On the Origin of Cattle: How Aurochs Became Cattle and Colonized the World." *Evolutionary Anthropology: Issues, News, and Reviews* 19, no. 4 (July 1, 2010): 148–57. doi:10.1002/evan.20267.

"The American Chestnut Research and Restoration Project." College of Environmental Science and Forestry (ESF), State University of New York. Accessed April 22, 2015. http://www.esf.edu/chestnut/#.VTdpThdPKIo.

"Antibodies Part 1: CRISPR." (Podcast.) *Radiolab.* Accessed January 16, 2016. http://www.radiolab.org/story/antibodies-part-1-crispr/.

Axelrod, Daniel I., and Harry P. Bailey. "Cretaceous Dinosaur Extinction." *Evolution* 22, no. 3 (September 1, 1968): 595–611. doi:10.2307/2406883.

Bambach, Richard K. "Phanerozoic Biodiversity Mass Extinctions." *Annual Review of Earth and Planetary Sciences* 34, no. 1 (2006): 127–55. doi:10.1146/annurev.earth.33.092203.122654.

Barnosky, Anthony D., Nicholas Matzke, Susumu Tomiya, Guinevere O.U. Wogan, Brian Swartz, Tiago B. Quental, Charles Marshall, et al. "Has the Earth's Sixth Mass Extinction Already Arrived?" *Nature* 471, no. 7336 (March 3, 2011): 51–57. doi:10.1038/nature09678.

Bennett, Joseph R., Richard F. Maloney, Tammy E. Steeves, James Brazill-Boast, Hugh P. Possingham, and Philip J. Seddon. "Spending Limited Resources on De-extinction Could Lead to Net Biodiversity Loss." *Nature Ecology & Evolution* 1 (March 1, 2017): 53. doi:10.1038/s41559-016-0053.

Bodmer, Sir Walter. *Successful Science Communication: Telling It Like It Is*. Edited by David J. Bennett and Richard C. Jennings. Cambridge; New York: Cambridge University Press, 2011.

Brand, Stewart. "Back from the Brink: Should We Use Cloning to Save Endangered Species?" *Slate*, June 4, 2014. http://www.slate.com/articles/technology/future_tense/2014/06/cloning_wildlife_carrie_friese_s_book_on_using_tech_to_save_endangered_species.html.

———. "Earth Monitoring: Whole Earth Comes into Focus." *Nature* 450, no. 7171 (December 6, 2007): 797. doi:10.1038/450797a.

———. *Whole Earth Discipline: An Ecopragmatist Manifesto*. New York: Viking, 2009.

Brandom, Russell. "Woolly Mammoth Rising: Can We Bring Extinct Animals Back to Life?" *The Verge*, March 18, 2013. http://www.theverge.com/2013/3/18/4119062/can-deextinction-bring-back-the-mammoth-passenger-pigeon-thylacine.

Buckley, Stephanie Gruner. "Blood and Spore: How a Bat-Killing Fungus Is Threatening U.S. Agriculture." *The Atlantic*, May 6, 2013. http://www.theatlantic.com/business/archive/2013/05/blood-and-spore-how-a-bat-killing-fungus-is-threatening-us-agriculture/275596/.

Bull, J.W., and M. Maron. "How Humans Drive Speciation as Well as Extinction." *Proceedings of the Royal Society B* 283, no. 1833 (June 29, 2016). doi:10.1098/rspb.2016.0600.

Caldeira, Ken, and Michael E. Wickett. "Oceanography: Anthropogenic Carbon and Ocean pH." *Nature* 425, no. 6956 (September 25, 2003): 365. doi:10.1038/425365a.

Callaway, Ewen. " 'Dino-Chickens' Reveal How the Beak Was Born." *Nature News*, May 12, 2015. doi:10.1038/nature.2015.17507.

———. "First Synthetic Yeast Chromosome Revealed." *Nature News*, March 27, 2014. doi:10.1038/nature.2014.14941.

———. "Mammoth Genomes Provide Recipe for Creating Arctic Elephants." *Nature* 521, no. 7550 (May 1, 2015): 18–19. doi:10.1038/nature.2015.17462.

———. "Stem-Cell Plan Aims to Bring Rhino Back from Brink of Extinction." *Nature* 533, no. 7601 (May 3, 2016): 20–21. doi:10.1038/533020a.

Camacho, Alejandro E. "Going the Way of the Dodo: De-extinction, Dualisms, and Reframing Conservation." SSRN Scholarly Paper. Rochester, NY: Social Science Research Network, August 11, 2014. http://papers.ssrn.com/abstract=2478815.

Campbell, Kevin L., Jason E.E. Roberts, Laura N. Watson, Jörg Stetefeld, Angela M. Sloan, Anthony V. Signore, Jesse W. Howatt, et al. "Substitutions in Woolly Mammoth Hemoglobin Confer Biochemical Properties Adaptive for Cold Tolerance." *Nature Genetics* 42, no. 6 (June 2010): 536–40. doi:10.1038/ng.574.

Cardillo, Marcel, Georgina M. Mace, Kate E. Jones, Jon Bielby, Olaf R.P. Bininda-Emonds, Wes Sechrest, C. David L. Orme, and Andy Purvis. "Multiple Causes of High Extinction Risk in Large Mammal Species." *Science* 309, no. 5738 (August 19, 2005): 1239–41. doi:10.1126/science.1116030.

Caro, Tim. "The Pleistocene Re-wilding Gambit." *Trends in Ecology & Evolution* 22, no. 6 (June 2007): 281–83. doi:10.1016/j.tree.2007.03.001.

Caro, Tim, Jack Darwin, Tavis Forrester, Cynthia Ledoux-Bloom, and Caitlin Wells. "Conservation in the Anthropocene." *Conservation Biology* 26, no. 1 (February 1, 2012): 185–88. doi:10.1111/j.1523-1739.2011.01752.x.

Ceballos, Gerardo, Paul R. Ehrlich, Anthony D. Barnosky, Andrés García, Robert M. Pringle, and Todd M. Palmer. "Accelerated Modern Human–Induced Species Losses: Entering the Sixth Mass Extinction." *Science Advances* 1, no. 5 (June 1, 2015). doi:10.1126/sciadv.1400253.

Charo, R. Alta, and Henry T. Greely. "CRISPR Critters and CRISPR Cracks." *The American Journal of Bioethics* 15, no. 12 (December 2, 2015): 11–17. doi:10.1080/15265161.2015.1104138.

Charpentier, Emmanuelle, and Jennifer A. Doudna. "Biotechnology: Rewriting a Genome." *Nature* 495, no. 7439 (March 7, 2013): 50–51. doi:10.1038/495050a.

Church, George. "George Church: De-extinction Is a Good Idea." *Scientific American,* September 1, 2013. http://www.scientificamerican.com/article/george-church-de-extinction-is-a-good-idea/.

Church, George M., and Ed Regis. *Regenesis: How Synthetic Biology Will Re-invent Nature and Ourselves.* New York: Basic Books, 2012.

Clubb, Ros, Marcus Rowcliffe, Phyllis Lee, Khyne U. Mar, Cynthia Moss, and Georgia J. Mason. "Compromised Survivorship in Zoo Elephants." *Science* 322, no. 5908 (December 12, 2008): 1649. doi:10.1126/science.1164298.

Colebrook, Claire. *The Death of the PostHuman: Essays on Extinction, Volume One.* Ann Arbor, MI: Michigan Publishing/Open Humanities Press, 2014. http://hdl.handle.net/2027/spo.12329362.0001.001.

———. *Sex after Life: Essays on Extinction, Volume Two.* Ann Arbor, MI: Michigan Publishing/Open Humanities Press, 2014. http://hdl.handle.net/2027/spo.12329363.0001.001.

Collard, Rosemary-Claire, Jessica Dempsey, and Juanita Sundberg. "A Manifesto for Abundant Futures." *Annals of the Association of American Geographers* 105, no. 2 (March 4, 2015): 322–30. doi:10.1080/00045608.2014.973007.

Connaughton, Maddison. "Zoos, Conservation and the Fight for De-extinction." *The Saturday Paper* (Carlton, Australia), October 31, 2015. https://www.thesaturdaypaper.com.au/2015/10/31/zoos-conservation-and-the-fight-de-extinction/14462100002554.

Corbyn, Zoë. "Can We Save the Rhino from Poachers with a 3D Printer?" *The Guardian,* May 24, 2015. http://www.theguardian.com/environment/2015/may/24/artificial-3d-printed-fake-rhino-horn-poaching.

Corlett, Richard T. "The Anthropocene Concept in Ecology and Conservation." *Trends in Ecology & Evolution* 30, no. 1 (January 2015): 36–41. doi:10.1016/j.tree.2014.10.007.

———. "A Bigger Toolbox: Biotechnology in Biodiversity Conservation." *Trends in Biotechnology* 35, no. 1 (July 13, 2016). doi:10.1016/j.tibtech.2016.06.009.

———. "The Role of Rewilding in Landscape Design for Conservation." *Current Landscape Ecology Reports* 1, no. 3 (August 24, 2016): 127–33. doi:10.1007/s40823-016-0014-9.

Cornell, Maraya. "Why Are Most of Tanzania's Elephants Disappearing?" *National Geographic*, June 12, 2015. http://news.nationalgeographic.com/2015/06/150612-tanzania-environmental-investigation-agency-mary-rice-elephants-poaching-cites-corruption/.

Crichton, Michael. *Jurassic Park: A Novel*. New York: Knopf, 1990.

Crutzen, Paul J. "Geology of Mankind." *Nature* 415, no. 6867 (January 3, 2002): 23. doi:10.1038/415023a.

Crutzen, Paul J., and Eugene F. Stoermer. "The 'Anthropocene.'" *Global Change Newsletter*, no. 41 (May 2000): 17–18.

Cyranoski, David. "Gene-Edited 'Micropigs' to Be Sold as Pets at Chinese Institute." *Nature* 526, no. 7571 (September 29, 2015): 18. doi:10.1038/nature.2015.18448.

Dannemann, Michael, Aida M. Andrés, and Janet Kelso. "Introgression of Neandertal- and Denisovan-like Haplotypes Contributes to Adaptive Variation in Human Toll-like Receptors." *American Journal of Human Genetics* 98, no. 1 (January 7, 2016): 22–33. doi:10.1016/j.ajhg.2015.11.015.

Darwin, Charles. *The Variation of Animals and Plants under Domestication*. New York: Orange Judd, 1868.

"De-extinction Debate: Should We Bring Back the Woolly Mammoth?" Point/Counterpoint, Yale Environment 360, January 13, 2014. http://e360.yale.edu/feature/the_case_for_de-extinction_why_we_should_bring_back_the_woolly_mammoth/2721/.

De Vos, Jurriaan M., Lucas N. Joppa, John L. Gittleman, Patrick R. Stephens, and Stuart L. Pimm. "Estimating the Normal Background Rate of Species Extinction." *Conservation Biology* 29, no. 2 (April 1, 2015): 452–62. doi:10.1111/cobi.12380.

Dewey, John. *The Public and Its Problems: An Essay in Political Inquiry*. Athens, OH: Swallow Press, 1954.

Donlan, C. Josh. "De-extinction in a Crisis Discipline." *Frontiers of Biogeography* 6, no. 1 (January 1, 2014). http://escholarship.org/uc/item/2x7oq4nk.

Donlan, C. Josh, Joel Berger, Carl E. Bock, Jane H. Bock, David A. Burney, James A. Estes, Dave Foreman, et al. "Pleistocene Rewilding: An Optimistic Agenda for Twenty-First Century Conservation." *The American Naturalist* 168, no. 5 (November 1, 2006): 660–81. doi:10.1086/508027.

Edwards, Ceiridwen J., David A. Magee, Stephen D.E. Park, Paul A. McGettigan, Amanda J. Lohan, Alison Murphy, Emma K. Finlay, et al. "A Complete Mitochondrial Genome

Sequence from a Mesolithic Wild Aurochs (*Bos primigenius*)." *PLOS ONE* 5, no. 2 (February 17, 2010). doi:10.1371/journal.pone.0009255.

Ehrenfeld, David. "Transgenics and Vertebrate Cloning as Tools for Species Conservation." *Conservation Biology* 20, no. 3 (June 2006): 723-32. doi:10.1111/j.1523-1739.2006.00399.x.

Eisenmann, V., and J.S. Brink. "Koffiefontein Quaggas and True Cape Quaggas: The Importance of Basic Skull Morphology." *South African Journal of Science* 96, no. 9/10 (2000): 529-33.

Epstein, Brendan, Menna Jones, Rodrigo Hamede, Sarah Hendricks, Hamish McCallum, Elizabeth P. Murchison, Barbara Schönfeld, Cody Wiench, Paul Hohenlohe, and Andrew Storfer. "Rapid Evolutionary Response to a Transmissible Cancer in Tasmanian Devils." *Nature Communications* 7 (August 30, 2016). doi:10.1038/ncomms12684.

Esvelt, Kevin M., Andrea L. Smidler, Flaminia Catteruccia, and George M. Church. "Concerning RNA-Guided Gene Drives for the Alteration of Wild Populations." *eLife,* July 17, 2014. doi:10.7554/eLife.03401.

"Extinction: Biology, Culture, and Our Futures." Symposium. Yale University, New Haven, CT, October 11, 2014. http://frankeprogram.yale.edu/event/extinction-biology-culture-and-our-futures.

"Extinction Marathon: Visions of the Future." Exhibition. Serpentine Sackler Gallery, London, October 18-19, 2014. Accessed November 2, 2014. http://www.serpentinegalleries.org/exhibitions-events/extinction-marathon.

Field, Dawn, and Neil Davies. *Biocode: The New Age of Genomics.* Oxford: Oxford University Press, 2015.

Flagg, Anna. "A Disappearing Planet." (Interactive infographic.) *ProPublica.* Accessed October 30, 2014. http://projects.propublica.org/extinctions.

Fletcher, Amy Lynn. "Bio-imaginaries: Bringing Back the Woolly Mammoth." Chapter 6 in *Mendel's Ark: Biotechnology and the Future of Extinction,* 89-99. New York: Springer, 2014. http://link.springer.com/chapter/10.1007/978-94-017-9121-2_6.

———. "Genuine Fakes: Cloning Extinct Species as Science and Spectacle." *Politics and the Life Sciences* 29, no. 1 (March 1, 2010): 48-60. doi:10.2990/29_1_48.

———. *Mendel's Ark: Biotechnology and the Future of Extinction.* New York: Springer, 2014.

Friese, Carrie. *Cloning Wild Life: Zoos, Captivity, and the Future of Endangered Animals.* New York: NYU Press, 2013.

Friese, Carrie, and Claire Marris. "Making De-extinction Mundane?" *PLOS Biology* 12, no. 3 (March 25, 2014). doi:10.1371/journal.pbio.1001825.

Fuller, Errol. *The Great Auk.* New York: Abrams, 1999.

Gantz, Valentino M., Nijole Jasinskiene, Olga Tatarenkova, Aniko Fazekas, Vanessa M. Macias, Ethan Bier, and Anthony A. James. "Highly Efficient Cas9-Mediated

Gene Drive for Population Modification of the Malaria Vector Mosquito *Anopheles stephensi.*" *Proceedings of the National Academy of Sciences* 112, no. 49 (November 23, 2015): E6736-43. doi:10.1073/pnas.1521077112.

Gazeau, Frédéric, Laura M. Parker, Steeve Comeau, Jean-Pierre Gattuso, Wayne A. O'Connor, Sophie Martin, Hans-Otto Pörtner, and Pauline M. Ross. "Impacts of Ocean Acidification on Marine Shelled Molluscs." *Marine Biology* 160, no. 8 (August 2013): 2207-45. doi:10.1007/s00227-013-2219-3.

Gibbons, Ann. "The Thoroughly Bred Horse." *Science* 346, no. 6216 (December 19, 2014): 1439. doi:10.1126/science.346.6216.1439.

Green, Richard E., Johannes Krause, Adrian W. Briggs, Tomislav Maricic, Udo Stenzel, Martin Kircher, Nick Patterson, et al. "A Draft Sequence of the Neandertal Genome." *Science* 328, no. 5979 (May 7, 2010): 710-22. doi:10.1126/science.1188021.

Greenberg, Joel. *A Feathered River Across the Sky: The Passenger Pigeon's Flight to Extinction.* New York: Bloomsbury, 2014.

Grimaldi, David A. *Amber: Window to the Past.* New York: Abrams; American Museum of Natural History, 1996.

Gunasena, K.T., J.R.T. Lakey, P.M. Villines, M. Bush, C. Raath, E.S. Critser, L.E. McGann, and J.K. Critser. "Antral Follicles Develop in Xenografted Cryopreserved African Elephant (*Loxodonta africana*) Ovarian Tissue." *Animal Reproduction Science* 53, no. 1-4 (October 30, 1998): 265-75. doi:10.1016/s0378-4320(98)00132-8.

Gurusamy, V., A. Tribe, and C.J.C. Phillips. "Identification of Major Welfare Issues for Captive Elephant Husbandry by Stakeholders." *Animal Welfare* 23, no. 1 (February 1, 2014): 11-24. doi:10.7120/09627286.23.1.011.

Guthrie, Richard. "The Catastrophic Nature of Humans." *Nature Geoscience* 8, no. 6 (June 2015): 421-22. doi:10.1038/ngeo2455.

Haraway, Donna Jeanne. *When Species Meet.* Minneapolis: University of Minnesota Press, 2007.

Hautmann, Michael. "Effect of End-Triassic CO2 Maximum on Carbonate Sedimentation and Marine Mass Extinction." *Facies* 50, no. 2 (July 28, 2004): 257-61. doi:10.1007/s10347-004-0020-y.

Heard, Matthew Joshua. "De-extinction: Raising the Dead and a Number of Important Questions." *Frontiers of Biogeography* 6, no. 1 (January 1, 2014). http://escholarship.org/uc/item/18h9k112.

Herridge, Tori. "Mammoths Are a Huge Part of My Life. But Cloning Them Is Wrong." *The Guardian,* November 18, 2014. http://www.theguardian.com/commentisfree/2014/nov/18/mammoth-cloning-wrong-save-endangered-elephants.

Higuchi, Russell, Barbara Bowman, Mary Freiberger, Oliver A. Ryder, and Allan C. Wilson. "DNA Sequences from the Quagga, an Extinct Member of the Horse Family." *Nature* 312, no. 5991 (November 15, 1984): 282-84. doi:10.1038/312282a0.

Higuchi, Russell G., Lisa A. Wrischnik, Elizabeth Oakes, Matthew George, Benton Tong, and Allan C. Wilson. "Mitochondrial DNA of the Extinct Quagga: Relatedness and

Extent of Postmortem Change." *Journal of Molecular Evolution* 25, no. 4 (1987): 283–87. doi:10.1007/BF02603111.

Holmberg, Tora. "Mortal Love: Care Practices in Animal Experimentation." *Feminist Theory* 12, no. 2 (August 1, 2011): 147–63. doi:10.1177/1464700111404206.

Holmes, R. Max. "An Arctic Solution to Climate Warming." TEDxWoods-Hole, November 11, 2011. https://www.youtube.com/watch?v=vBrFudwiVpo.

Hoshino, Yoichiro, Noboru Hayashi, Shunji Taniguchi, Naohiko Kobayashi, Kenji Sakai, Tsuyoshi Otani, Akira Iritani, and Kazuhiro Saeki. "Resurrection of a Bull by Cloning from Organs Frozen without Cryoprotectant in a –80°C Freezer for a Decade." *PLOS ONE* 4, no. 1 (January 8, 2009). doi:10.1371/journal.pone.0004142.

Hung, Chih-Ming, Pei-Jen L. Shaner, Robert M. Zink, Wei-Chung Liu, Te-Chin Chu, Wen-San Huang, and Shou-Hsien Li. "Drastic Population Fluctuations Explain the Rapid Extinction of the Passenger Pigeon." *Proceedings of the National Academy of Sciences* 111, no. 29 (July 22, 2014): 10636–41. doi:10.1073/pnas.1401526111.

"Interview with George Church: Can Neanderthals Be Brought Back from the Dead?" *Spiegel Online*, January 18, 2013. http://www.spiegel.de/international/zeitgeist/george-church-explains-how-dna-will-be-construction-material-of-the-future-a-877634.html.

Jablonski, David. "Background and Mass Extinctions: The Alternation of Macroevolutionary Regimes." *Science* 231, no. 4734 (January 10, 1986): 129–33. doi:10.1126/science.231.4734.129.

———. "Extinction: Past and Present." *Nature* 427, no. 6975 (February 12, 2004): 589. doi:10.1038/427589a.

———. "Lessons from the Past: Evolutionary Impacts of Mass Extinctions." *Proceedings of the National Academy of Sciences* 98, no. 10 (May 8, 2001): 5393–98. doi:10.1073/pnas.101092598.

James, David N. "Ectogenesis: A Reply to Singer and Wells." *Bioethics* 1, no. 1 (January 1987): 80–99. doi:10.1111/j.1467-8519.1987.tb00006.x.

Jones, Kate Elizabeth. "From Dinosaurs to Dodos: Who Could and Should We De-extinct?" *Frontiers of Biogeography* 6, no. 1 (January 1, 2014). http://escholarship.org/uc/item/9gv7n6d3.

Jurassic Park. Directed by Steven Spielberg. Screenplay by Michael Crichton and David Koepp. Los Angeles: Universal Pictures, 1993.

Kaebnick, Gregory E. *Humans in Nature: The World as We Find It and the World as We Create It*. New York: Oxford University Press, 2013.

Kato, Hiromi, Masayuki Anzai, Tasuku Mitani, Masahiro Morita, Yui Nishiyama, Akemi Nakao, Kenji Kondo, et al. "Recovery of Cell Nuclei from 15,000 Years Old Mammoth Tissues and Its Injection into Mouse Enucleated Matured Oocytes." *Proceedings of the Japan Academy. Series B, Physical and Biological Sciences* 85, no. 7 (July 2009): 240–47. doi:10.2183/pjab.85.240.

Kjelland, Michael E., Salvador Romo, and Duane C. Kraemer. "Avian Cloning: Adaptation of a Technique for Enucleation of the Avian Ovum." *Avian Biology Research* 7, no. 3 (2014): 131–38. doi:10.3184/175815514X14065426847114.

Klass, Perri. "The Artificial Womb Is Born." *New York Times Magazine,* September 29, 1996. http://www.nytimes.com/1996/09/29/magazine/the-artificial-womb-is-born.html.

Klein, Richard G., and Kathryn Cruz-Uribe. "Craniometry of the Genus *Equus* and the Taxonomic Affinities of the Extinct South African Quagga." *South African Journal of Science* 95, no. 2 (1999): 81–86.

Knoepfler, Paul. GMO *Sapiens: The Life-Changing Science of Designer Babies.* Hackensack, NJ: World Scientific, 2015.

Kolbert, Elizabeth. "Martha, My Dear: What De-extinction Can't Bring Back." *The New Yorker,* March 11, 2014. http://www.newyorker.com/news/daily-comment/martha-my-dear-what-de-extinction-cant-bring-back.

———. "Recall of the Wild." *The New Yorker,* December 24 and 31, 2012. http://www.newyorker.com/magazine/2012/12/24/recall-of-the-wild.

———. *The Sixth Extinction: An Unnatural History.* New York: Henry Holt, 2014.

Landecker, Hannah. "Living Differently in Time: Plasticity, Temporality and Cellular Biotechnologies." *Culture Machine* 7 (2005). http://www.culturemachine.net/index.php/cm/article/view/26.

Lander, Eric S. "The Heroes of CRISPR." *Cell* 164, no. 1 (January 14, 2016): 18–28. doi:10.1016/j.cell.2015.12.041.

Lanphier, Edward, Fyodor Urnov, Sarah Ehlen Haecker, Michael Werner, and Joanna Smolenski. "Don't Edit the Human Germ Line." *Nature* 519, no. 7544 (March 12, 2015): 410–11. doi:10.1038/519410a.

Lanza, Robert P., Jose B. Cibelli, Francisca Diaz, Carlos T. Moraes, Peter W. Farin, Charlotte E. Farin, Carolyn J. Hammer, et al. "Cloning of an Endangered Species (*Bos gaurus*) Using Interspecies Nuclear Transfer." *Cloning* 2, no. 2 (2000): 79–90. doi:10.1089/152045500436104.

Latour, Bruno. " 'Fifty Shades of Green': Bruno Latour on the Ecomodernist Manifesto." Presentation to the Panel on Modernism, Breakthrough Dialogue, Sausalito, CA, June 2015. Entitle Blog, June 27, 2015. https://entitleblog.org/2015/06/27/fifty-shades-of-green-bruno-latour-on-the-ecomodernist-manifesto/.

Ledford, Heidi. "CRISPR, the Disruptor." *Nature* 522, no. 7554 (June 3, 2015): 20–24. doi:10.1038/522020a.

Ledford, Heidi, and Ewen Callaway. " 'Gene Drive' Mosquitoes Engineered to Fight Malaria." *Nature News,* November 23, 2015. doi:10.1038/nature.2015.18858.

Liang, Puping, Yanwen Xu, Xiya Zhang, Chenhui Ding, Rui Huang, Zhen Zhang, Jie Lv, et al. "CRISPR/Cas9-Mediated Gene Editing in Human Tripronuclear Zygotes." *Protein & Cell* 6, no. 5 (April 18, 2015): 363–72. doi:10.1007/s13238-015-0153-5.

Lynch, Vincent J., Oscar C. Bedoya-Reina, Aakrosh Ratan, Michael Sulak, Daniela I. Drautz-Moses, George H. Perry, Webb Miller, and Stephan C. Schuster. "Elephantid Genomes Reveal the Molecular Bases of Woolly Mammoth Adaptations to the Arctic." *Cell Reports* 12, no. 2 (July 14, 2015): 217–28. Published online July 2, 2015. doi:10.1016/j.celrep.2015.06.027.

Marshall, Michael. "Neanderthal–Human Sex Bred Light Skins and Infertility." *New Scientist,* January 29, 2014. https://www.newscientist.com/article/mg22129542-600-neanderthal-human-sex-bred-light-skins-and-infertility/.

Martinelli, Lucia, Markku Oksanen, and Helena Siipi. "De-extinction: A Novel and Remarkable Case of Bio-objectification." *Croatian Medical Journal* 55, no. 4 (August 2014): 423–27. doi:10.3325/cmj.2014.55.423.

Maxmen, Amy. "The Genesis Engine." *WIRED,* July 22, 2015. http://www.wired.com/2015/07/crispr-dna-editing-2/.

McKie, Robin. "The Quest Is to Clone a Mammoth. The Question Is: Should We Do It?" *The Guardian,* July 14, 2013. http://www.theguardian.com/science/2013/jul/14/wooly-mammoth-extinct-cloning-dna.

Meier, Rudolf, and Quentin D. Wheeler, eds. *Species Concepts and Phylogenetic Theory: A Debate.* New York: Columbia University Press, 2000.

Miller, Webb, Daniela I. Drautz, Aakrosh Ratan, Barbara Pusey, Ji Qi, Arthur M. Lesk, Lynn P. Tomsho, et al. "Sequencing the Nuclear Genome of the Extinct Woolly Mammoth." *Nature* 456, no. 7220 (November 20, 2008): 387–90. doi:10.1038/nature07446.

Mills, L. Scott, Michael E. Soulé, and Daniel F. Doak. "The Keystone-Species Concept in Ecology and Conservation." *BioScience* 43, no. 4 (April 1, 1993): 219–24. doi:10.2307/1312122.

Minteer, Ben A. "Extinct Species Should Stay Extinct." *Slate,* December 1, 2014. http://www.slate.com/articles/technology/future_tense/2014/12/de_extinction_ethics_why_extinct_species_shouldn_t_be_brought_back.html.

———. "Is It Right to Reverse Extinction?" *Nature* 509, no. 7500 (May 14, 2014): 261. doi:10.1038/509261a.

Monbiot, George. *Feral: Rewilding the Land, Sea and Human Life.* London: Penguin, 2014.

Mooallem, Jon. *Wild Ones: A Sometimes Dismaying, Weirdly Reassuring Story about Looking at People Looking at Animals in America.* New York: Penguin, 2013.

Morton, Timothy. *Hyperobjects: Philosophy and Ecology after the End of the World.* Minneapolis: University of Minnesota Press, 2013.

Newhouse, Andrew E., Linda D. Polin-McGuigan, Kathleen A. Baier, Kristia E.R. Valletta, William H. Rottmann, Timothy J. Tschaplinski, Charles A. Maynard, and William A. Powell. "Transgenic American Chestnuts Show Enhanced Blight Resistance and Transmit the Trait to T1 Progeny." *Plant Science* 228 (November 2014): 88–97. Published online April 13, 2014. doi:10.1016/j.plantsci.2014.04.004.

Nicholls, Henry. "Darwin 200: Let's Make a Mammoth." *Nature* 456, no. 7220 (November 20, 2008): 310–14. doi:10.1038/456310a.

O'Connor, M.R. "Making Rhino Horns out of Stem Cells." *The Atlantic*, December 24, 2014. http://www.theatlantic.com/technology/archive/2014/12/making-rhino-horns-from-stem-cells/384039/.

——. *Resurrection Science: Conservation, De-extinction and the Precarious Future of Wild Things*. New York: St. Martin's Press, 2015.

Oksanen, Markku, and Helena Siipi, eds. *The Ethics of Animal Re-creation and Modification: Reviving, Rewilding, Restoring*. Basingstoke, UK: Palgrave Macmillan, 2014.

Oye, Kenneth A., Kevin Esvelt, Evan Appleton, Flaminia Catteruccia, George Church, Todd Kuiken, Shlomiya Bar-Yam Lightfoot, et al. "Regulating Gene Drives." *Science* 345, no. 6197 (August 8, 2014): 626–28. doi:10.1126/science.1254287.

Pääbo, Svante. "Neanderthals Are People, Too." *New York Times*, April 24, 2014. http://www.nytimes.com/2014/04/25/opinion/neanderthals-are-people-too.html.

Paddle, Robert. *The Last Tasmanian Tiger: The History and Extinction of the Thylacine*. Cambridge: Cambridge University Press, 2002.

Padian, Kevin, and Luis M. Chiappe. "The Origin of Birds and Their Flight." *Scientific American* 278, no. 2 (February 1, 1998): 28–37. https://www.scientificamerican.com/article/the-origin-of-birds-and-their-fligh/.

Palkopoulou, Eleftheria, Swapan Mallick, Pontus Skoglund, Jacob Enk, Nadin Rohland, Heng Li, Ayça Omrak, et al. "Complete Genomes Reveal Signatures of Demographic and Genetic Declines in the Woolly Mammoth." *Current Biology* 25, no. 10 (May 18, 2015): 1395–1400. Published online April 23, 2015. doi:10.1016/j.cub.2015.04.007.

Peers, Michael J.L., Daniel H. Thornton, Yasmine N. Majchrzak, Guillaume Bastille-Rousseau, and Dennis L. Murray. "De-extinction Potential under Climate Change: Extensive Mismatch between Historic and Future Habitat Suitability for Three Candidate Birds." *Biological Conservation* 197 (May 2016): 164–70. doi:10.1016/j.biocon.2016.03.003.

Penney, David, Caroline Wadsworth, Graeme Fox, Sandra L. Kennedy, Richard Preziosi, and Terence Brown. "Absence of Ancient DNA in Sub-fossil Insect Inclusions Preserved in 'Anthropocene' Colombian Copal." *PLOS ONE* 8, no. 9 (September 11, 2013). doi:10.1371/journal.pone.0073150.

Pilcher, Helen. *Bring Back the King: The New Science of De-extinction*. London: Bloomsbury Sigma, 2016.

Pimm, L.L., L. Dollar, and O.L. Bass. "The Genetic Rescue of the Florida Panther." *Animal Conservation* 9, no. 2 (May 1, 2006): 115–22. doi:10.1111/j.1469-1795.2005.00010.x.

Pimm, Stuart. "Opinion: The Case against Species Revival." *National Geographic News*, March 12, 2013. http://news.nationalgeographic.com/news/2013/03/130312--deextinction-conservation-animals-science-extinction-biodiversity-habitat-environment.

Pimm, S.L., C.N. Jenkins, R. Abell, T.M. Brooks, J.L. Gittleman, L.N. Joppa, P.H. Raven, C.M. Roberts, and J.O. Sexton. "The Biodiversity of Species and Their Rates of Extinction, Distribution, and Protection." *Science* 344, no. 6187 (May 30, 2014). doi:10.1126/science.1246752.

Poinar, George, Jr. "Ancient DNA: The Tools of Molecular Biology Can Be Used to Peer into an Organism's Genetic Future—or Its Distant Past." *American Scientist* 87, no. 5 (September 1, 1999): 446–57. http://www.jstor.org/stable/27857905.

Poinar, George O., Jr., and Roberta Hess. "Ultrastructure of 40-Million-Year-Old Insect Tissue." *Science* 215, no. 4537 (March 5, 1982): 1241–42. doi:10.1126/science.215. 4537.1241.

Poinar, Hendrik N., Carsten Schwarz, Ji Qi, Beth Shapiro, Ross D.E. MacPhee, Bernard Buigues, Alexei Tikhonov, et al. "Metagenomics to Paleogenomics: Large-Scale Sequencing of Mammoth DNA." *Science* 311, no. 5759 (January 20, 2006): 392–94. doi:10.1126/science.1123360.

Pollack, Andrew. "Concerns Are Raised about Genetically Engineered Mosquitoes." *New York Times*, October 30, 2011. http://www.nytimes.com/2011/10/31/science/ concerns-raised-about-genetically-engineered-mosquitoes.html.

———. "A Powerful New Way to Edit DNA." *New York Times*, March 3, 2014. http://www. nytimes.com/2014/03/04/health/a-powerful-new-way-to-edit-dna.html.

Powell, Alvin. "Behold the Mammoth (Maybe)." *Harvard Gazette*, October 14, 2014. http://news.harvard.edu/gazette/story/2014/10/behold-the-mammoth-maybe/.

Reardon, Sara. "Gene-Editing Record Smashed in Pigs." *Nature*, October 6, 2015. doi:10.1038/nature.2015.18525.

Redford, Kent H., William Adams, Rob Carlson, Georgina M. Mace, and Bertina Ceccarelli. "Synthetic Biology and the Conservation of Biodiversity." *Oryx* 48, no. 3 (July 2014): 330–36. doi:10.1017/S0030605314000040.

"Rhino." World Wildlife Fund. Accessed October 6, 2015. http://www.worldwildlife.org/ species/rhino.

Rich, Nathaniel. "The Mammoth Cometh." *New York Times Magazine*, February 27, 2014. http://www.nytimes.com/2014/03/02/magazine/the-mammoth-cometh.html.

Richmond, Douglas J., Mikkel-Holger S. Sinding, and M. Thomas P. Gilbert. "The Potential and Pitfalls of De-extinction." *Zoologica Scripta* 45, suppl. s1 (October 1, 2016): 22–36. doi:10.1111/zsc.12212.

Robin, Libby. "Histories for Changing Times: Entering the Anthropocene?" *Australian Historical Studies* 44, no. 3 (September 2013): 329–40. doi:10.1080/1031461X. 2013.817455.

Sandler, Ronald. "The Ethics of Reviving Long Extinct Species." *Conservation Biology* 28, no. 2 (April 1, 2014): 354–60. doi:10.1111/cobi.12198.

Sankararaman, Sriram, Swapan Mallick, Michael Dannemann, Kay Prüfer, Janet Kelso, Svante Pääbo, Nick Patterson, and David Reich. "The Genomic Landscape of Neanderthal Ancestry in Present-Day Humans." *Nature* 507, no. 7492 (January 29, 2014): 354–57. doi:10.1038/nature12961.

Santymire, Rachel M., Travis M. Livieri, Heather Branvold-Faber, and Paul E. Marinari. "The Black-Footed Ferret: On the Brink of Recovery?" In *Reproductive Sciences in Animal Conservation,* edited by William V. Holt, Janine L. Brown, and Pierre Comizzoli, 119–34. Advances in Experimental Medicine and Biology 753. New York: Springer, 2014. doi:10.1007/978-1-4939-0820-2_7.

Sartore, Joel. "Photo Ark." *National Geographic.* Accessed February 3, 2017. http://www. nationalgeographic.org/projects/photo-ark/.

Schorger, Arlie William. *The Passenger Pigeon, Its Natural History and Extinction.* Madison: University of Wisconsin Press, 1955.

Schubert, Mikkel, Hákon Jónsson, Dan Chang, Clio Der Sarkissian, Luca Ermini, Aurélien Ginolhac, Anders Albrechtsen, et al. "Prehistoric Genomes Reveal the Genetic Foundation and Cost of Horse Domestication." *Proceedings of the National Academy of Sciences* 111, no. 52 (December 30, 2014): E5661–69. doi:10.1073/pnas.1416991111.

Seddon, Philip J. "De-extinction: Reframing the Possible." *Trends in Ecology & Evolution* 30, no. 10 (January 10, 2015): 569–70. doi:10.1016/j.tree.2015.08.002.

——. "From Reintroduction to Assisted Colonization: Moving along the Conservation Translocation Spectrum." *Restoration Ecology* 18, no. 6 (November 1, 2010): 796–802. doi:10.1111/j.1526-100X.2010.00724.x.

Seddon, Philip J., Christine J. Griffiths, Pritpal S. Soorae, and Doug P. Armstrong. "Reversing Defaunation: Restoring Species in a Changing World." *Science* 345, no. 6195 (July 25, 2014): 406–12. doi:10.1126/science.1251818.

Seddon, Philip J., Axel Moehrenschlager, and John Ewen. "Reintroducing Resurrected Species: Selecting DeExtinction Candidates." *Trends in Ecology & Evolution* 29, no. 3 (March 2014): 140–47. doi:10.1016/j.tree.2014.01.007.

Sepkoski, J.J., Jr. "Phanerozoic Overview of Mass Extinction." In *Patterns and Processes in the History of Life,* edited by D. M. Raup and D. Jablonski, 277–95. Dahlem Workshop Reports 36. Berlin; Heidelberg: Springer, 1986.

Shapiro, Beth. "Could We 'De-extinctify' the Woolly Mammoth?" *The Guardian,* April 26, 2015. http://www.theguardian.com/science/2015/apr/26/woolly-mammoth-normal-for-norfolk-de-extinction.

——. *How to Clone a Mammoth: The Science of De-extinction.* Princeton, NJ: Princeton University Press, 2015.

Shapiro, Beth, Alexei J. Drummond, Andrew Rambaut, Michael C. Wilson, Paul E. Matheus, Andrei V. Sher, Oliver G. Pybus, et al. "Rise and Fall of the Beringian Steppe Bison." *Science* 306, no. 5701 (November 26, 2004): 1561–65. doi:10.1126/science.1101074.

Shapiro, B., and M. Hofreiter. "A Paleogenomic Perspective on Evolution and Gene Function: New Insights from Ancient DNA." *Science* 343, no. 6169 (January 24, 2014). doi:10.1126/science.1236573.

Shapiro, Beth, Dean Sibthorpe, Andrew Rambaut, Jeremy Austin, Graham M. Wragg, Olaf R. P. Bininda-Emonds, Patricia L.M. Lee, and Alan Cooper. "Flight of the Dodo." *Science* 295, no. 5560 (March 1, 2002): 1683. doi:10.1126/science.295.5560.1683.

Sheehan, Peter M. "The Late Ordovician Mass Extinction." *Annual Review of Earth and Planetary Sciences* 29, no. 1 (2001): 331–64. doi:10.1146/annurev.earth.29.1.331.

Shelley, Mary. *Frankenstein: or, the Modern Prometheus: The Original 1818 Text.* 2nd edition. Peterborough, ON: Broadview Press, 1999.

Sherkow, J.S., and H.T. Greely. "What If Extinction Is Not Forever?" *Science* 340, no. 6128 (April 5, 2013): 32–33. doi:10.1126/science.1236965.

Silver, Lee M. *Challenging Nature: The Clash of Science and Spirituality at the New Frontiers of Life.* New York: Ecco, 2006.

Snyder, Gary. *The Practice of the Wild.* Berkeley, CA: Counterpoint Press, 2010.

Soares, André E. R., Ben J. Novak, James Haile, Tim H. Heupink, Jon Fjeldså, M. Thomas P. Gilbert, Hendrik Poinar, George M. Church, and Beth Shapiro. "Complete Mitochondrial Genomes of Living and Extinct Pigeons Revise the Timing of the Columbiform Radiation." *BMC Evolutionary Biology* 16 (2016): 230. doi:10.1186/s12862-016-0800-3.

Steyn, Paul. "Largest Wildlife Census in History Makes Waves in Conservation." *National Geographic,* January 4, 2016. http://news.nationalgeographic.com/2016/01/160104-great-elephant-census-vulcan-paul-allen-elephants-conservation/.

St. Fleur, Nicholas. "The Genetically Modified Mosquito Bite." *The Atlantic,* January 27, 2015. http://www.theatlantic.com/health/archive/2015/01/Genetically-Modified-Mosquitoes-May-Be-Released-in-Florida-Keys/384859/.

Stilgoe, Jack. "*Jurassic World: Frankenstein* for the 21st Century?" *The Guardian,* June 24, 2015. http://www.theguardian.com/film/political-science/2015/jun/24/jurassic-world-frankenstein-for-the-21st-century.

Stinson, Liz. "To Save Our Ecosystems, Will We Have to Design Synthetic Creatures?" *WIRED,* January 7, 2015. http://www.wired.com/2015/01/save-ecosystems-will-design-synthetic-creatures/.

Stokstad, Erik. "Bringing Back the Aurochs." *Science* 350, no. 6265 (December 4, 2015): 1144–47. doi:10.1126/science.350.6265.1144.

Stone, R. "A Rescue Mission for Amphibians at the Brink of Extinction." *Science* 339, no. 6126 (March 22, 2013): 1371. doi:10.1126/science.339.6126.1371.

Swart, Sandra. "Frankenzebra: Dangerous Knowledge and the Narrative Construction of Monsters." *Journal of Literary Studies* 30, no. 4 (October 2014): 45–70. doi:10.1080/02564718.2014.976456.

Tarnocai, C., J.G. Canadell, E.A.G. Schuur, P. Kuhry, G. Mazhitova, and S. Zimov. "Soil Organic Carbon Pools in the Northern Circumpolar Permafrost Region." *Global Biogeochemical Cycles* 23, no. 2 (June 27, 2009). doi:10.1029/2008GB003327.

Taylor, Victoria J., and Trevor B. Poole. "Captive Breeding and Infant Mortality in Asian Elephants: A Comparison between Twenty Western Zoos and Three Eastern Elephant Centers." *Zoo Biology* 17, no. 4 (December 7, 1998): 311–32. doi:10.1002/(SICI)1098-2361(1998)17:4<311::AID-ZOO5>3.0.CO;2-C.

TEDxDeExtinction. Washington, DC, March 15, 2013. Accessed January 18, 2015. https://www.ted.com/tedx/events/7650.

Thomas, Chris D. "The Anthropocene Could Raise Biological Diversity." *Nature* 502, no. 7469 (October 2, 2013): 7. doi:10.1038/502007a.

Thompson, Helen. "Plant Science: The Chestnut Resurrection." *Nature* 490, no. 7418 (October 3, 2012): 22–23. doi:10.1038/490022a.

Turner, Derek D. "De-extinction as Artificial Species Selection." *Philosophy & Technology,* September 18, 2016. doi:10.1007/s13347-016-0232-4.

Turner, Fred. *From Counterculture to Cyberculture: Stewart Brand, the Whole Earth Network, and the Rise of Digital Utopianism.* Chicago: University of Chicago Press, 2010.

Tyack, Scott G., Kristie A. Jenkins, Terri E. O'Neil, Terry G. Wise, Kirsten R. Morris, Matthew P. Bruce, Scott McLeod, et al. "A New Method for Producing Transgenic Birds via Direct in Vivo Transfection of Primordial Germ Cells." *Transgenic Research* 22, no. 6 (December 2013): 1257–64. doi:10.1007/s11248-013-9727-2.

Vajta, Gábor, and Mickey Gjerris. "Science and Technology of Farm Animal Cloning: State of the Art." *Animal Reproduction Science* 92, nos. 3–4 (May 2006): 211–30. doi:10.1016/j.anireprosci.2005.12.001.

van Dooren, Thom. *Flight Ways: Life and Loss at the Edge of Extinction.* New York: Columbia University Press, 2014.

van Dooren, Thom, and Deborah Bird Rose. "Keeping Faith with the Dead: Mourning and De-extinction." *Australian Zoologist,* June 3, 2015. doi:10.7882/AZ.2014.048.

Velasquez-Manoff, Moises. "Should You Fear the Pizzly Bear?" *New York Times Magazine,* August 14, 2014. http://www.nytimes.com/2014/08/17/magazine/should-you-fear-the-pizzly-bear.html.

Vernot, B., and J.M. Akey. "Resurrecting Surviving Neandertal Lineages from Modern Human Genomes." *Science* 343, no. 6174 (February 28, 2014): 1017–21. doi:10.1126/science.1245938.

Vogel, Gretchen. "Cloned Gaur a Short-Lived Success." *Science* 291, no. 5503 (January 19, 2001): 409. doi:10.1126/science.291.5503.409A.

Wade, Nicholas. "Regenerating a Mammoth for $10 Million." *New York Times,* November 20, 2008. http://www.nytimes.com/2008/11/20/science/20mammoth.html.

———. "Scientists Seek Ban on Method of Editing the Human Genome." *New York Times,* March 19, 2015. http://www.nytimes.com/2015/03/20/science/biologists-call-for-halt-to-gene-editing-technique-in-humans.html.

Whittle, Patrick M., Emma J. Stewart, and David Fisher. "Re-creation Tourism: De-extinction and Its Implications for Nature-Based Recreation." *Current Issues in Tourism* 18, no. 10 (April 15, 2015): 1–5. doi:10.1080/13683500.2015.1031727.

Woese, Carl R. "A New Biology for a New Century." *Microbiology and Molecular Biology Reviews* 68, no. 2 (June 1, 2004): 173–86. doi:10.1128/MMBR.68.2.173-186.2004.

Wroe, Stephen, Philip Clausen, Colin McHenry, Karen Moreno, and Eleanor Cunningham. "Computer Simulation of Feeding Behaviour in the Thylacine and Dingo as a Novel Test for Convergence and Niche Overlap." *Proceedings of the Royal Society of London B: Biological Sciences* 274, no. 1627 (November 22, 2007): 2819–28. doi:10.1098/rspb.2007.0906.

Yin, Hao, Wen Xue, Sidi Chen, Roman L. Bogorad, Eric Benedetti, Markus Grompe, Victor Koteliansky, et al. "Genome Editing with Cas9 in Adult Mice Corrects a Disease Mutation and Phenotype." *Nature Biotechnology* 32, no. 6 (June 2014): 551–53. doi:10.1038/nbt.2884.

Yong, Ed. "Will We Ever… Clone a Mammoth?" BBC Future, June 1, 2012. http://www.bbc.com/future/story/20120601-will-we-ever-clone-a-mammoth.

Zalasiewicz, Jan, Mark Williams, Alan Smith, Tiffany L. Barry, Angela L. Coe, Paul R. Bown, Patrick Brenchley, et al. "Are We Now Living in the Anthropocene?" *GSA Today* (Geological Society of America) 18, no. 2 (2008): 4. doi:10.1130/GSAT01802A.1.

Zalasiewicz, Jan, Mark Williams, Will Steffen, and Paul Crutzen. "The New World of the Anthropocene." *Environmental Science & Technology* 44, no. 7 (April 2010): 2228–31. doi:10.1021/es903118j.

Zimmer, Carl. "Bringing Them Back to Life: The Revival of an Extinct Species Is No Longer a Fantasy. But Is It a Good Idea?" *National Geographic*, April 2013. http://ngm.nationalgeographic.com/2013/04/125-species-revival/zimmer-text.

Zimov, Sergey A. "Pleistocene Park: Return of the Mammoth's Ecosystem." *Science* 308, no. 5728 (May 6, 2005), 796–98. doi:10.1126/science.1113442.

Zimov, S.A., S.P. Davydov, G.M. Zimova, A.I. Davydova, E.A.G. Schuur, K. Dutta, and F.S. Chapin. "Permafrost Carbon: Stock and Decomposability of a Globally Significant Carbon Pool." *Geophysical Research Letters* 33, no. 20 (October 1, 2006). doi:10.1029/2006GL027484.

Zimov, Sergey A., Edward A.G. Schuur, and F. Stuart Chapin. "Permafrost and the Global Carbon Budget." *Science* 312, no. 5780 (June 16, 2006): 1612–13. doi:10.1126/science.1128908.

Zimov, S.A., N.S. Zimov, A.N. Tikhonov, and F.S. Chapin III. "Mammoth Steppe: A High-Productivity Phenomenon." *Quaternary Science Reviews* 57 (December 4, 2012): 26–45. doi:10.1016/j.quascirev.2012.10.005.

Zylinska, Joanna. *Minimal Ethics for the Anthropocene*. Ann Arbor, MI: Michigan Publishing/Open Humanities Press, 2014. http://hdl.handle.net/2027/spo.12917741.0001.001.

INDEX

Abumrad, Jad, 219
African wildcat, 43
Allee effect, 173-75
amber, DNA extraction from, 19-23
American chestnut, 221-23
American Museum of Natural History, 205
Ansel, W.F.A., 32
Anthropocene: critique of, 67; definition, 66-67; evolutionary adaption to, 68-69; extinction rate, 67-68
archaea, 50
Archer, Michael, 85, 87, 90-91, 92, 229
artificial selection, 29. *See also* selective breeding
artificial wombs, 144
Asimov, Isaac, 231-32
Attig, Thomas, 241-42
auk, great, 15-16, 182-84, 191
aurochs, 16, 17, 33, 36-39, 100
Australian Museum, 233

backbreeding (breeding back), 35, 37
bacteria, 49, 50
band-tailed pigeon, 152, 164-65, 166-67, 170, 174
banteng, 43, 196
bats, 203, 209
beavers, 83, 94, 95
Beijing Genomics Institute, 52
ben-Aaron, Diana, 134-35
Bennett, Joseph, 74-75
Berg, Paul, 47-48
Bering Land Bridge, 126-27
Bible, 228, 229
bioabundance, 69, 70
biological species concept, 26-27
biotechnology: societal acceptance of, 223-24, 226-27. *See also* CRISPR; gene drive; gene editing; genetic

engineering; genetic rescue; Revive & Restore; selective breeding
birds, 168-69, 180-81, 182. See also *specific species*
bison, 118, 120-21, 185
black-footed ferret, 207-9, 220
blight, chestnut, 221-22
borders, international, 196
Boyer, Herbert, 48
Brand, Stewart: on adaptation in Anthropocene, 68-69; background, 4; on CRISPR, 53; on current extinction discussions, 149; eco-pragmatism of, 245; on gene drive, 220-21; on *Jurassic Park*, 14; on mass extinction lens, 69-70; on moral imperative of de-extinction, 14; on razorbills and great auks, 184; reintroduction focus, 80; on taking action, 244-45; *Whole Earth Catalogue*, 4-5. *See also* Phelan, Ryan; Revive & Restore
breeding back (backbreeding), 35, 37
bucardo (Pyrenean ibex), 17, 45-46, 187, 205
business opportunities, 37, 39, 100, 110-11, 189-90, 233. *See also* entertainment; funding

Campbell, Kevin, 137
carbon, in permafrost, 114-17, 147
care, excessive, 234-35
Carlin, Norman, 187-88, 189-90, 195, 198
Carolina parakeet, 175, 190, 248
Cartagena Protocol on Biosafety, 196
Cas9 enzyme, 50, 184-85. *See also* CRISPR
cattle, 36-37. *See also* aurochs
cautionary vigilance, 4
cave paintings, 36, 126
cells, 39-40, 138-40

DAVID SUZUKI INSTITUTE

THE DAVID SUZUKI INSTITUTE is a non-profit organization founded in 2010 to stimulate debate and action on environmental issues. The Institute and the David Suzuki Foundation both work to advance awareness of environmental issues important to all Canadians.

We invite you to support the activities of the Institute. For more information please contact us at:

David Suzuki Institute
219–2211 West 4th Avenue
Vancouver, BC, Canada V6K 4S2
info@davidsuzukiinstitute.org
604-742-2899
www.davidsuzukiinstitute.org

Cheques can be made payable to The David Suzuki Institute.